NATIONAL CENTER FOR EDUCATION STATISTICS

Technical Report December 1994

National Survey of Recent College Graduates

A Study of Selected Nonsampling Errors in the 1991 Survey of Recent College Graduates

J. Michael Brick
Margaret Cahalan
Lucinda Gray
Jacqueline Severynse
Westat, Inc.

Peter Stowe
Project Officer
National Center for Education Statistics

U.S. Department of Education
Office of Educational Research and Improvement NCES 95-640

U.S. Department of Education
Richard W. Riley
Secretary

Office of Educational Research and Improvement
Sharon P. Robinson
Assistant Secretary

National Center for Education Statistics
Emerson J. Elliott
Commissioner

National Center for Education Statistics

"The purpose of the Center shall be to collect, analyze, and disseminate statistics and other data related to education in the United States and in other nations."—Section 406(b) of the General Education Provisions Act, as amended (20 U.S.C. 1221e–1).

December 1994

Contact:
Peter Stowe
(202) 219–1363

For sale by the U.S. Government Printing Office
Superintendent of Documents, Mail Stop: SSOP, Washington, DC 20402-9328
ISBN 0-16-045434-4

HIGHLIGHTS

The three major sources of nonsampling error in the RCG:91 are errors due to nonresponse, random measurement errors, and systematic errors due to interviewers. Errors of these types exist in the estimators from virtually every survey, but often there is little evidence that can be used to quantify nonsampling errors. In the RCG:91, special efforts were made to measure the nonsampling errors. The potential consequences for users of the RCG:91 data and suggested areas for improvement in the future studies based on these efforts are presented. Some of the major findings are given below.

- Sampled units that do not participate in a survey are a source of bias. In the RCG:91 the institution response rate was 95 percent and the graduate response rate was 83 percent. The overall two-stage response rate was 79 percent. While these rates are higher than previous RCG surveys, the bias due to nonresponse is still an important component of error in the RCG:91.

- As with most surveys, the nonresponse bias in the RCG:91 is likely to be more significant for estimates based on large sample sizes, especially when the characteristic is highly correlated with the response rate.

- Statistical adjustments were made to reduce the bias due to nonresponse. This evaluation of the estimates shows that the estimates were subject to relatively small biases due to nonresponse.

- The biases in the estimates were also computed from reinterviews. These data show the biases are generally small and not statistically significant. The response variances computed from the reinterview data are typically moderate and the ordinary estimates of sampling errors account for these types of nonsampling errors. The estimates from the reinterview are valuable for improving the questionnaires for future surveys.

- A third assessment of errors was done to examine the contribution of interviewers to the errors in the estimates. The systematic errors associated with interviewers are very small, but the effects on the errors of the estimates could still be important for some types of estimates. The effects are most important for questions asked of all or almost all sampled graduates.

- Weighting adjustments beyond those already included in the final survey weights to account for nonresponse and response bias are not recommended. The adjustment of the standard errors of the estimates to account for measurement error introduced by interviewers is feasible, but not generally recommended. These adjustments are small or moderate for many estimates.

- Conservative inference procedures, such as using 99 percent confidence intervals in place of 95 percent intervals, are one way of protecting users from making erroneous inferences for the survey estimates. These methods increase the probability of preparing confidence intervals that cover the population value. This method can be used in addition to the procedure for adjusting the standard errors of the estimates.

ACKNOWLEDGMENTS

Many individuals made significant contributions to the 1991 Survey of Recent College Graduates and to the accompanying evaluation studies. The authors gratefully acknowledge their efforts. The survey was performed under the direction of the National Center for Education Statistics (NCES) Postsecondary Division Cross-Sectional Studies Branch. Paul Planchon was the Associate Commissioner for the Postsecondary Division and Roslyn Korb was the Cross-Sectional Studies branch director. Peter Stowe was the NCES project officer and Michael Cohen was the NCES mathematical statistician.

The survey was performed under contract with Westat, Inc. The Westat project team included Margaret Cahalan, project director; Lucinda Gray, survey manager; Mike Brick, senior statistician; Jacqueline Severynse, statistician; Peter Ha, Gail Wisan, and Steven Schweinfurth, analysis and sampling programming; Susan Hein, graphics; Sylvie Warren, word processing; Carol Litman, editor; Jacque Wernimont, Royce Gibson, and Nancy Hopper, CATI development; Karen Molloy, Telephone Research Center coordinator; and Stephanie Campbell and Dotty Pike, data preparation. The study benefitted from the corporate support and encouragement of Westat vice president Lance Hodes.

Critical technical review of this report was provided by NCES staff Bob Burton, Jim Houser, and Steve Kaufman. John Bushery of the U.S. Census Bureau also provided technical review. The authors wish to thank each of these individuals for their careful reading of this report and their helpful comments and suggestions.

The authors especially acknowledge with gratitude the 400 higher education institutions that provided information necessary to draw the sample of graduates, the 14,000 graduates who took time to respond to the survey, and the 10 state certification agencies participating in the validity study. Together these groups provided the information upon which this report is based.

CONTENTS

CONTENTS -- Continued

CONTENTS - Continued

LIST OF APPENDICES

LIST OF TABLES

CONTENTS - Continued

LIST OF TABLES - Continued

LIST OF FIGURES

CONTENTS - Continued

LIST OF FIGURES - Continued

LIST OF EXHIBITS

1. INTRODUCTION

The 1991 Survey of Recent College Graduates (RCG:91) is the sixth study in a series of surveys begun in 1976 by the National Center for Education Statistics (NCES). The series provides data on the occupational and educational outcomes of recent bachelor's and master's graduates 1 year after graduation. Survey information was collected on graduates' labor force status, occupation, relationship of employment to major field of study, enrollment since graduation, and teacher qualification status. The RCG:91 was conducted by Westat, Inc., for the NCES.

The RCG:91 was a two-stage sample. A sample of 400 higher education institutions awarding baccalaureate degrees was selected in the first stage, and a sample of 18,000 persons who received bachelor's and master's degrees in 1989-90 was selected from these sampled institutions in the second stage. Data were collected by means of computer assisted telephone interviewing (CATI) from July 1991 to December 1991. To be included in the survey, graduates had to meet the following criteria: (1) they received a bachelor's or master's degree from the college or university from which they were sampled; (2) they received their degree between July 1, 1989, and June 30, 1990; and (3) they lived in the United States at the time of the survey.[1] The weighted response rates for schools was 95 percent, and the weighted response rate for graduates was 83 percent.

The estimates from the RCG surveys are used to prepare reports, such as the "Occupational and Educational Outcomes of Recent College Graduates 1 Year after Graduation: 1991" (Cahalan et al., 1993). Statements of the reliability and accuracy of statistics in these reports recognize that the estimates are subject to variation from two major sources. One is sampling error due to sampling and interviewing only a fraction of the institutions and graduates in the population. The other source of error is generically called nonsampling error, and it includes all errors that are not due to sampling. The sources of nonsampling errors include errors due to incomplete responses, ambiguity in the meaning of the questions, interviewer errors, respondent errors, processing errors, and incomplete lists used to survey the target population, to name just a few.

Along with the estimates, reports typically include the standard errors of the estimates. These standard errors provide users with information on the nature and size of the sampling error. The standard errors can also be used to make inferences from the data, such as confidence intervals and tests of significance. However, the errors due to nonsampling error are often not included in the estimated standard errors. This point is discussed more fully in the body of this report.

---

[1]Respondents who were out of the country for the entire data collection period (July to December 1991) were excluded from the study.

This report examines nonsampling errors and their impact on the estimates from the RCG:91. The goal is to inform users of the potential for error and how these errors may influence inferential statements. Another important goal of this report is to identify specific procedures, such as questionnaire construction and data collection methods, that are likely to contribute to the errors in the estimates. Recommendations for users of the RCG:91 estimates and for designers of subsequent surveys in the series are also presented.

NCES STANDARDS

In January 1992, NCES adopted a set of standards that apply to all of the work conducted by and for the Center (Flemming, 1992). One of these standards (V-01-92) pertains to the evaluation of surveys. The statement of the purpose of this standard is

> The results of the statistical evaluation must enable users of the survey data to understand the quality and limitations of the data and must provide information for planning future surveys or replications of the same survey. Also the inclusion of a systematic assessment of all sources of nonsampling error for key statistics to be studied or reported in NCES publications.

The goals of this report support the purpose embodied in this standard. The type of systematic assessment of the sources of nonsampling error is only feasible when there are data to support it. In the RCG:91, resources were committed to this assessment.

RCG: 91 ASSESSMENT STUDIES

For RCG:91 four assessments of nonsampling errors were conducted: an analysis of nonresponse; an analysis of measurement error by means of a reinterview; an analysis of the impact of contribution of interviewers to nonsampling error; and an analysis of the validity of data on teacher certification by an administrative records check. A brief introduction to each of these studies is given below.

- **Nonresponse analysis.** Characteristics of respondents and nonrespondents are compared to assess the potential nonsampling error due to nonresponse. The analysis concentrates on sampled graduates who did not respond to the survey, since this type of nonresponse has the greatest potential for influencing the estimates from the survey. Nonresponse from institutions that did not participate and nonresponse for specific items from the participating graduates are discussed briefly.

- **Reinterview analysis.** A sample of graduates who responded to the main survey was selected and these graduates were interviewed a second time. The data from the reinterview are compared with the responses from the original interview to estimate the potential for systematic and random measurement errors in the survey estimates.

- **Interviewer error analysis.** Completed interviews in the main survey are identified by interviewer and analyzed to assess the potential for additional errors due to the specific methods of the interviewer. These interviewer-level differences contribute to additional errors in the estimates.

- **Validity study analysis.** Data from state certification agencies were collected for a sample of graduates who reported in the survey that they were certified to teach. The certification data from graduates are compared to certification data reported by the state agencies. These data provide an estimate of the bias and random measurement errors from the survey.

Coverage and Other Nonsampling Errors

One source of nonsampling error that is often important in sample surveys is that due to incomplete coverage of the target population, in this case all bachelor's and master's degree recipients in the 1989-90 school year. In the RCG:91, coverage errors could result from either the sampling frame of institutions being incomplete or from the failure to include all the graduates when the list of graduates within the sampled institutions was created.

Coverage errors were not considered in this report for two main reasons. First, the coverage of graduates is believed to be very complete and not a large contributor to errors in the estimates. The sampling frame of institutions was the Integrated Postsecondary Education Data System (IPEDS) and its coverage of institutions awarding bachelor's and master's degrees is very complete. Peng (1979) discusses the quality of these data. The list of graduates from the sampled institutions were also thought to be very complete, especially since checks of these counts were built into the data collection process.

The second reason that coverage errors were not considered is that no data on coverage were collected to use for assessment. Without this type of data, the evaluation of coverage errors would be very speculative.

Other sources of nonsampling error could also have been considered in this report. For example, the coding of a few of the open-ended responses is a source of nonsampling error that could have been considered. The reasons given above for not assessing coverage error also apply to these other sources of nonsampling error. The report focuses on those nonsampling errors that are likely to have the most substantial impact on the estimates from the survey.

Method of Analysis

Nonsampling errors can be studied using a variety of methods. The analytic method used in this report is to develop models of the nonsampling errors and then use the data from the various assessment studies to estimate the parameters of these models and the impact on the estimates. This approach was chosen because it requires an explicit declaration of the assumptions of the model and it can be applied across a variety of sources of nonsampling error.

While this modeling approach does not result in a completely unified approach, it is more coherent than other approaches that were considered. Other approaches to the assessment might lead to the use of a variety of different statistical tools. For example, correlations are often used to evaluate the reliability of responses in education measurement studies. These measures are valid for some purposes, but they cannot be easily applied to the models to estimate the impact of the errors on the estimates. They also tend to make the evaluation of systematic errors distinct from random errors, even though the two are highly related.

One of the consequences of choosing the model approach to study nonsampling errors is that the report contains a significant amount of technical statistical concepts. These discussions are needed to adequately describe the models, to justify the methods used to estimate the parameters of the model, and to apply the methods to the RCG:91 data. In most cases, the technical detail is supplemented with a definition of the terms and heuristic explanations, where this is possible.

STRUCTURE OF THE REPORT

The report presents the results of each of the analyses in a separate chapter. Each chapter includes a section that describes the implications of the findings for users of the data and recommendations for future surveys. Nonresponse and its impact on the estimates is the topic of Chapter 2. In Chapter 3, measurement error is modeled using the data from the reinterview of the graduates. Both systematic and random measurement errors are considered. Chapter 4 extends the model from the previous chapter to include the contribution of interviewers to measurement error. Chapter 5 contains a discussion of nonsampling errors for teacher certification issues using data from the state certification agencies.

The last chapter attempts to integrate the findings from the earlier chapters. It begins by comparing the results from the reinterview and validity studies of teacher certification. A more comprehensive model of nonsampling errors is then discussed and some examples using estimates from the RCG:91 are presented. The chapter ends with general recommendations for users of the data and designers of future studies.

The report also contains a number of appendices with more detailed data on specific topics. The contents of these appendices are essential to support and justify some of the recommendations given in the body of the report.

2. NONSAMPLING ERROR FROM NONRESPONSE

One of the most pervasive and challenging sources of nonsampling error in estimates from sample surveys is the bias associated with nonresponse. Nonresponse bias can arise when a response is not obtained for a sampled unit or when a response is missing for an item in an otherwise completed interview.

Nonresponse bias is a function of both the amount of incompleteness and the difference in the characteristics between respondents and nonrespondents. A more detailed explanation of the relationship of these factors to the level of nonresponse bias is given in the next section. However, this relationship is the reason that both response rates and the difference in characteristics of respondents and nonrespondents need to be considered.

One of the reasons that it is so hard to evaluate nonresponse bias in the estimates from a survey is the lack of data for nonrespondents, which is critical in the evaluation. In most cases, only limited data are available for nonrespondents, and those data are usually restricted to a few characteristics from the frame from which the sample was selected. In some studies, a special study of nonrespondents is conducted to collect data for this type of evaluation, but the RCG:91 did not contain an intensive followup study of nonrespondents.

As a result, the estimation of nonresponse bias presented below is limited to a few variables for which data were collected for all sampled graduates (gender, degree, major, race and ethnicity, school control, and school size). The data for these items are available because the sampled institutions provided these data when they submitted the lists of graduates for sampling. These estimates of nonresponse bias are primarily indicators of the relative magnitude of the potential biases from this source. More than this is not possible without additional data collection from the nonrespondents.

The bias arising from nonresponse is of particular concern in surveys of recent college graduates because graduates typically move just following graduation and the information obtained from the institutions does not usually contain address updates. RCG studies prior to 1991 employed a mail data collection mode with telephone followup for a subsample of nonrespondents and, in the 1980s, generally achieved effective graduate response rates between 75 and 80 percent (Figure 2-1). The RCG:91 differed from previous data collections in that most interviews were completed by telephone using computer assisted telephone interviewing (CATI), and there was no subsampling of nonrespondents. A mail survey was employed only for those without telephone numbers and those who refused the telephone interview. The 83 percent graduate response rate for the 1991 survey suggests that these methods resulted in relatively high response rates for this part of the study without subsampling nonrespondents.

Figure 2-1. **RCG graduate response rates* over the last decade: 1981-91**

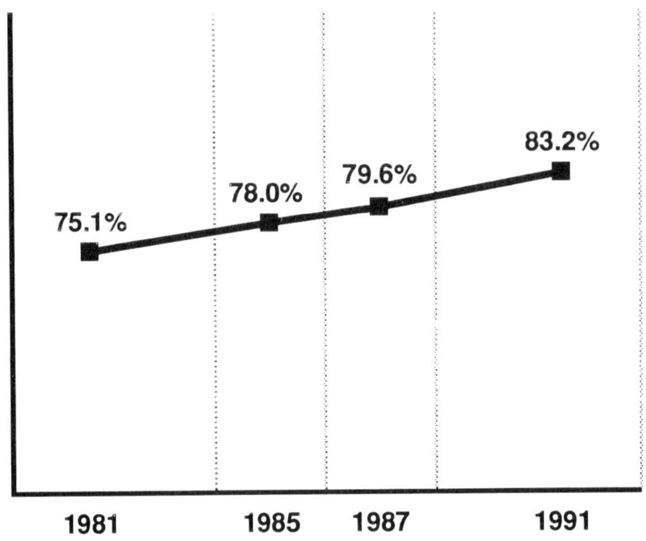

*1981 and 1985 rates are effective response rates, based on subsample of nonrespondents.

The next sections describe unit nonresponse and the potential for bias from this source. We briefly review the response rates and the characteristics of nonrespondents. We then develop a model for assessing nonresponse bias and apply this model to a few statistics from the RCG:91. Implications for data users and recommendations for future studies are then discussed. Item nonresponse and the implications for nonresponse bias from this source are presented at the end of this chapter.

UNIT RESPONSE RATES

The sample of graduates for the RCG:91 was obtained in two stages. First, a sample of 400 institutions awarding bachelor's or master's degrees was selected. Next, a sample of 18,135 graduates was selected from within the sampled institutions. In order to sample graduates, lists of all bachelor's and master's degree recipients from July 1, 1989, through June 30, 1990, were requested from each of the 400 sampled schools.

Unit nonresponse resulted if either the institution failed to cooperate with the survey or the graduate did not respond to the survey. The **unit response rate** is defined as the weighted number of eligible respondents divided by the weighted number of sampled units minus the weighted number of ineligible units. Graduate lists were obtained from 95 percent of the sampled schools. The institution response rates by control and size of the institution are given in Table 2-1 and displayed in Figure 2-2.

The overall response rate is the product of the institution response rate and the graduate response rate. Thus, the overall response rate for the RCG:91 was 79 percent (.79 = .95 x .83). In other words, interviews were not collected from approximately 20 percent of the graduates due to both institution and graduate nonresponse.

Table 2-1. Number of sampled institutions and weighted response rates, by institution control and size

Institution characteristic	Number of sampled institutions by status				Weighted response rate[2]
	Total	Participating	Nonresponse	Ineligible[1]	
Total	400	378	20	2	95.0%
Control					
Public	259	250	9	-	97.7
Private	141	128	11	2	93.6
Enrollment size					
Less than 1,500	189	179	8	2	95.2
1,500 - 5,999 .	191	180	11	-	93.4
6,000 or more .	20	19	1	-	95.0

[1]Of the two ineligibles, one school was closed and one had merged with another sampled institution.

[2]The weighted response rate is the weighted number of participating institutions divided by the sum of the weighted number of participants and nonrespondents.

SOURCE: U.S. Department of Education, National Center for Education Statistics, *1991 Recent College Graduates Survey.*

Figure 2-2. Institutional response rates by institution control: 1991, RCG

Graduate level nonresponse was a particular concern in the RCG:91. Because of the mobility of this population after graduation, there was the potential for substantial nonresponse due to not locating the sampled graduates. To address this issue, a number of tracing procedures were used to locate graduates to be interviewed. Some of these procedures were conducted prior to survey data collection (e.g., flyer mailing and post office updates), but most were conducted during data collection (e.g., alumni office requests, referrals, leads, and credit bureau searches). Once data collection began, 36 percent of the sample required tracing. Of the cases that required tracing, 72 percent were located. Details on these locating activities are included in Appendix A.

The graduate response rates for the RCG:91 are shown in Table 2-2. Of the sample of 18,135 graduates, 14,405 completed questionnaires, and 800 were ineligible for the survey. The unweighted response rate and the weighted (taking into consideration the unequal probabilities of selecting the graduates) response rate were both 83 percent.

Response rates by a few of the characteristics of sampled graduates are also shown in Table 2-2. For bachelor's degree recipients the weighted response rate was 84 percent, and for master's degree recipients it was 82 percent. Women responded at about the same rate (84 percent) as men (83 percent). The graduate characteristic with the greatest variation in response rates was race/ethnicity. White, non-Hispanic bachelor's graduates had the highest response rate (87 percent), and black, non-Hispanic graduates had the lowest rate (71 percent) (Figure 2-3).[2]

Figure 2-3. Bachelor's graduate response rate by race/ethnicity: 1991, RCG

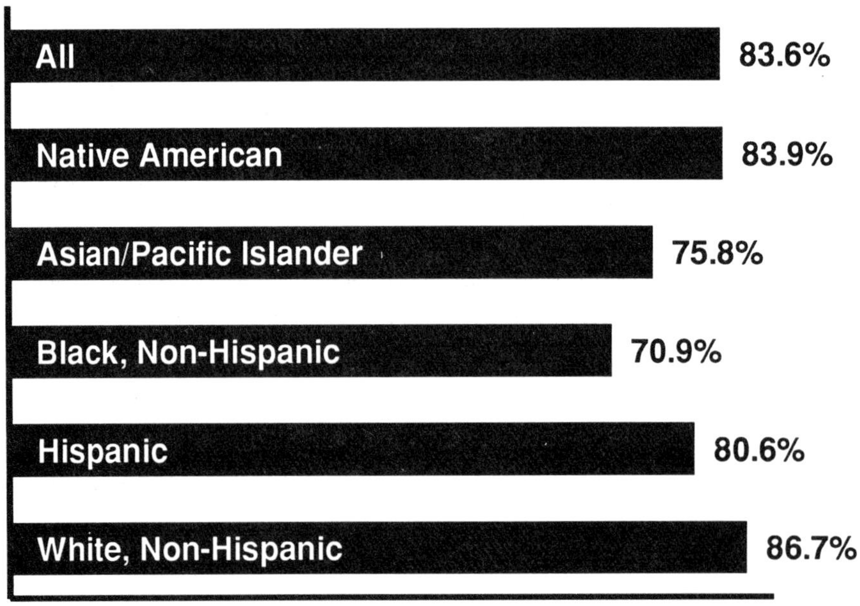

[2]The race/ethnicity codes are those reported by institutions for all bachelor's degree recipients and may not match data reported by the respondent on the survey. These items were collected from institutions for bachelor's degree recipients only, since they were not needed for sampling master's degree recipients. Of the graduates the institution identified as black, 97 percent also identified themselves as black on the survey. Of those the institution identified as Hispanic, 93 percent identified themselves as Hispanic on the survey.

Table 2-2. Number of sampled graduates and weighted response rates, by graduate and institution characteristics

Graduate and institution characteristic	Number of sampled graduates by status				Weighted response rate[2]
	Total	Complete	Nonresponse	Ineligible[1]	
Total	18,135	14,405	2,930	800	83.2%
Degree[3]					
Bachelor's	16,172	12,898	2,608	666	83.6
Master's	1,963	1,507	322	134	82.0
Major for bachelor's degree recipients[4]					
Education	3,109	2,630	381	98	87.3
Mathematics	379	325	43	11	87.8
Physical science	388	316	55	17	85.6
Other	12,296	9,627	2,129	540	83.1
Institution control					
Public	12,340	9,794	2,027	519	82.8
Private	5,795	4,611	903	281	84.1
Gender[5]					
Male	7,568	6,236	1,332	-	82.8
Female	9,767	8,169	1,598	-	83.7
Not coded	800	-	-	800	-
Race/ethnicity for bachelor's degree recipients[4]					
Native American	38	31	6	1	83.9
Asian or Pacific Islander	386	270	85	31	75.8
Black, non-Hispanic	1,743	1,187	484	72	70.9
Hispanic	709	544	128	37	80.6
White, non-Hispanic	8,803	7,425	1,076	302	86.7
Not reported[6]	4,493	3,441	829	223	80.2
Institution size					
Enrollment less than 1,500	7,617	6,134	1,170	313	84.6
Enrollment 1,500-5,999	8,549	6,715	1,441	393	82.4
Enrollment 6,000 or more	1,969	1,556	319	94	82.2

[1]The 800 ineligibles include graduates that did not receive their degree within the time frame (375), those living outside the country (368), those that received a degree other than bachelor's or master's (27), those deceased or incapacitated (25), and duplicates (5).

[2]The weighted response rate is the weighted number of completes divided by the sum of the weighted number of completes and nonrespondents.

[3]The degree codes are those reported by institutions for the entire sample and may not match data reported by the respondents on the survey.

[4]The major and race/ethnicity codes are those reported by institutions for all bachelor's degree recipients and may not match data reported by the respondents on the survey. These items were collected from institutions for bachelor's degree recipients only, since they were not needed for sampling master's degree recipients. Therefore, the columns for major and race/ethnicity will sum to the bachelor's degree totals. Of the graduates the institution identified as black, 97 percent also identified themselves as black on the survey. Of those the institution identified as Hispanic, 93 percent identified themselves as Hispanic on the survey.

[5]For respondents, the gender code was taken from the survey data. For nonrespondents, the gender was coded from the name. For ineligibles, the gender was not coded, since it was not needed to calculate response rates.

[6]Race/ethnicity was reported by about 72 percent of the institutions. Of the sampled graduates, 64 percent had race/ethnicity identified prior to sampling.

SOURCE: U.S. Department of Education, National Center for Education Statistics, *1991 Recent College Graduates Survey.*

Reasons for Nonresponse

The higher nonresponse rate among black graduates was related to greater difficulty in locating black graduates rather than higher refusal rates. In fact, refusal rates for blacks in the RCG:91 were actually slightly less than for whites (4.4 percent compared with 5.4 percent, Figure 2-4). This finding is consistent with other surveys that found lower or equal refusal rates for blacks (Groves, 1989; Weaver, Holmes, and Glenn, 1975; DeMaio, 1980), but found it more difficult to locate black respondents (Weaver, Holmes, and Glenn, 1975; Temple University, 1986-87).

Figure 2-4. Percentage of sample refusing interview by race/ethnicity: 1991, RCG

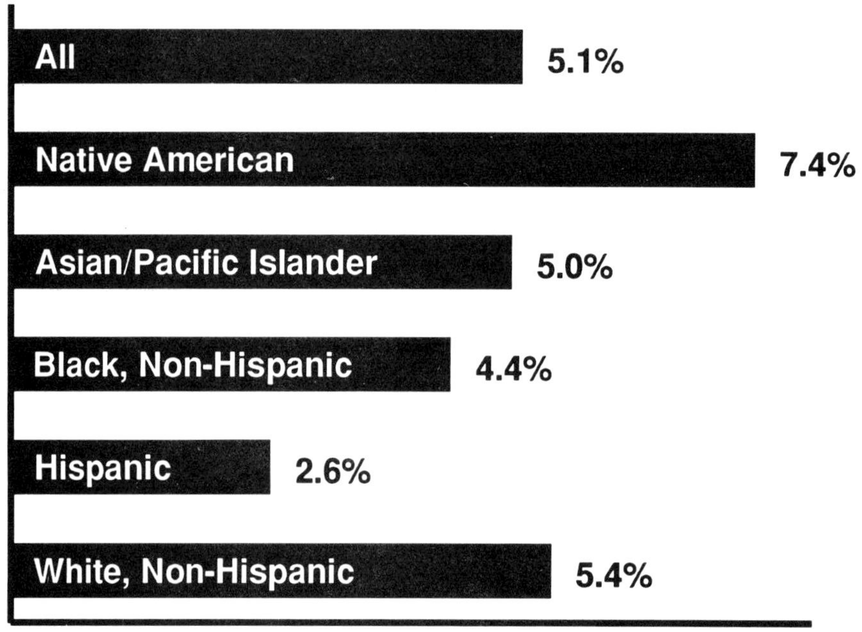

NOTE: Represents percent of total sample who were contacted by telephone and refused to do the interview.

Because the characteristics of respondents and nonrespondents may be different depending on the reason for nonresponse, the nonrespondents were classified by the reason for the nonresponse: refusal, nonlocatable, and other nonresponse (nonrespondents that were locatable but not available). The percent of nonrespondents by reason are shown in Table 2-3. Despite the success of the tracing operations, the main cause of nonresponse was still the inability to locate the sampled graduate. Of all the nonrespondents, 62 percent could not be located, 30 percent refused to participate, and 8 percent could not be interviewed after repeated telephone contacts to their households.

Table 2-3. Graduate nonresponse rates, by type of nonrespondent and characteristic of graduate

Graduate characteristic	Nonresponse rate	Percent due to		
		Refusal	Nonlocatable	Other
Total	16.8%	30.1%	61.6%	8.3%
Degree				
Bachelor's	16.4	31.1	60.5	8.4
Master's	18.0	26.9	65.0	8.1
Major*				
Education	12.7	31.9	59.6	8.5
Math	12.2	32.6	57.0	10.4
Physical Sciences	14.4	18.4	76.0	5.6
Other	16.9	31.2	60.4	8.4
Institution control				
Public	17.2	28.0	64.0	8.1
Private	15.9	34.8	56.3	8.9
Gender				
Male	17.2	28.4	63.6	8.1
Female	16.3	31.7	59.8	8.5
Race/ethnicity*				
Native American	16.1	46.0	54.0	0.0
Asian/Pacific Islander	24.2	20.8	72.2	7.0
Black, non-Hispanic	29.2	14.9	77.9	7.2
Hispanic	19.4	13.2	82.2	4.6
White, non-Hispanic	13.3	40.5	51.2	8.3
Not reported	19.8	24.6	66.3	9.1
Institution size class				
<1,500	15.4	29.2	62.5	8.3
1,500-5,999	17.6	31.1	61.1	7.8
6,000+	17.8	28.7	61.3	10.0

*Major and race/ethnicity were not collected for master's degree graduates.

NOTE: Percentages may not add to 100 due to rounding.

SOURCE: U.S. Department of Education, National Center for Education Statistics, *1991 Recent College Graduates Survey.*

Institution and Graduate Weight Adjustment

Weighting adjustments in the RCG survey were developed to partially address the potential nonresponse bias. In particular, nonresponse and poststratification adjustments were implemented to reduce the bias in the estimates. These adjustments took advantage of available data, especially data that were known to be related to nonresponse. Thus, characteristics such as the control of the institution and the race and ethnicity of the graduate were used to reduce the bias due to nonresponse.

In addition to the institution and graduate base weights, two institution-level adjustments, a graduate-level adjustment, and a poststratification adjustment were applied to reduce the bias in the estimates and variance. Specifically, the survey weights included the following components (described in more detail in the RCG:91 methodology report):

a. Institution base weight, the inverse of the probability of selection for the sampled institution.

b. Institution nonresponse adjustment to adjust for institutions that did not provide graduate lists. This used four institution categories based on control (public/private) and the emphasis in the programs of the school (bilingual education or other).

c. Institution-level ratio adjustment for the number of black and Hispanic graduates in the sampled schools.

d. Graduate base weight, the inverse of the probability of selection for the sampled graduate within the institution.

e. Graduate nonresponse adjustment to adjust for graduates that did not respond, using seven categories based on degree, race/ethnicity, and major field of study.

f. Poststratification adjustment using 20 categories based on respondent reported degree, major field of study, gender, and institution control.

Below, a model for evaluating the magnitude of the nonresponse bias is developed. Because of the potential importance of the weighting adjustments in reducing this source of bias, these estimation steps are included in the assessment.

MODEL OF GRADUATE NONRESPONSE BIAS

The result of not having complete data from all the sampled units is the potential for nonresponse bias. In this section, we examine the potential nonresponse bias arising from being unable to obtain the responses of sampled graduates. We exclude institutional nonresponse, primarily because only 5 percent of the sampled institutions did not respond and the bias from this source is likely to be relatively small.

We begin with a simple model for graduate nonresponse bias, then extend it to incorporate other significant features of the study. In the discussion, the focus is on bias despite the introduction of random error due to nonresponse. The contribution of random error due to nonresponse is included in the estimated standard error of the estimate, since the variance estimation method involved replicating the nonresponse adjustments. As a result, the standard error of the estimates can be used to estimate both the sampling error and the random error due to nonresponse.

The **nonresponse bias** of a linear estimate, such as a total, mean, or percent, can be defined as:

$$Bias(y_r) = \left(\frac{n-r}{n}\right)E(y_r - y_{nr}), \tag{2.1}$$

where

y_r	= estimate based on the r respondent cases;
y_{nr}	= estimate based on the $(n - r)$ nonrespondent cases;
n	= total sample size;
r	= number of respondents.

The operator, E, refers to the expectation over all possible samples. This expression is similar to the one proposed by Groves (1989). It helps clarify the relationship mentioned earlier between the response rate and the difference in the characteristics between respondents and nonrespondents. In other words, nonresponse bias is the product of the nonresponse rate and the difference between the respondent and nonrespondent estimates.

As formulated above, the bias of the estimate cannot be computed from any specific sample because it relies on averaging over all possible samples. However, we can use (2.1) as a model and estimate the bias using data on some variables that are available for all sampled graduates. This procedure conditions on the selected sample rather than averaging over all possible samples. As noted earlier, the lack of data on nonrespondents significantly limits the ability to use this model for a variety of statistics of interest in the RCG:91, but it at least provides an indication of the magnitude of the bias for a few statistics.

When using (2.1) as a model, we must recognize the sample design for the RCG:91 is not a simple random sample. Thus, the appropriately weighted estimates of the quantities given in the expression for the nonresponse bias must be substituted to account for the sample design. The weight is the inverse of the probability of selection of the graduate, including the probability of selection of the institution, the institution-level adjustments (nonresponse and ratio adjustments), and the probability of selection of the graduate from the list provided by the institution. Thus, the weight is the fully adjusted weight except it does not include the graduate nonresponse adjustment and the poststratification adjustment (includes components a-d described on page 2-8).

Estimates of Nonresponse Bias

The following example illustrates how the nonresponse bias is estimated using this model for a particular statistic. The estimate used in the example is the percentage of graduates with bachelor's degrees who are education majors. The estimates of bias are also expressed in different ways in these examples to illustrate the importance of the estimates.

Example. The weighted nonresponse rate for bachelor's graduates is 16.4 percent, and this value is used to estimate the first quantity in expression (2.1). The second quantity is the difference between the estimated proportion of respondents and nonrespondents who are education majors. Data on the frame (the list of graduates supplied by the sampled institutions) for the major of the sampled graduates was used to estimate the percentage of both nonrespondents who are education majors ($\hat{p}_{nr}$=7.8%) and respondents who are education majors ($\hat{p}_r$=10.5%). The bias for this estimate is modeled as follows:

$$\begin{aligned} \text{bias}(\hat{p}) &= (.164) * (.1054 - .0779) \\ &= 0.00451 \\ &= 0.5\% \end{aligned}$$

In other words, if the estimate were based only on the respondents it would overestimate the percentage who are education majors by 0.5 percent.

The **relative bias** is defined as the bias of the estimate divided by the estimate. Estimates of the relative bias indicate the order of magnitude of the bias with respect to the estimate. The relative bias may be of value to generalize the results to characteristics that cannot be modeled directly, due to the lack of data on nonrespondents.

In our example for the percentage of education majors, the relative bias is:

$$\begin{aligned} \text{rel bias}(\hat{p}) &= .00451/.1054 \\ &= .0428 \\ &= 4.3\% \end{aligned}$$

In this case the estimated bias in the percentage of education majors is small (less than 5 percent of the estimate).

The **bias ratio** is defined as the ratio of the bias of an estimate to its standard error. The bias ratio is another useful indicator of the impact of nonresponse bias. We will follow the general convention of expressing this ratio as a percentage. To understand why the bias ratio is important, consider the estimation confidence intervals or tests of significance.

In general, confidence intervals are not affected very much if the ratio of the bias to the standard error is less than 10 percent. For example, if the bias ratio is 10 percent, then the probability of an error of more than 1.96 standard deviations from the mean is 5.1 percent rather than the nominal 5 percent. As the bias ratio increases, the level of the confidence interval diverges more from the nominal level. When the bias equals the standard error (the bias ratio is 100 percent), the actual confidence interval is only 83 percent rather than the nominal 95 percent, as shown in the following table taken from Cochran (1977).

Bias ratio	Probability of an Error (Type I)
2%	0.0500
4%	0.0502
6%	0.0504
8%	0.0508
10%	0.0511
20%	0.0546
40%	0.0685
60%	0.0921
80%	0.1259
100%	0.1700
150%	0.3231

Continuing the example using the percentage of graduates who are education majors, the bias ratio is 148 percent. From the table above, the bias can be seen to have an important impact on the probability of an error. Instead of the nominal level of 5 percent, the probability increases to 32 percent when the bias ratio is this large. The bias ratio is so large in this case because the estimated standard error of the number of education majors is very small for this characteristic.

Extended Model for Major Components of Nonresponse

Groves (1989) pointed out that expressions like (2.1) do not adequately represent the various sources of nonresponse bias. Since the bias may be different depending on the reason for the nonresponse, he suggested expanding the expression to include the major sources of nonresponse. For the RCG:91, an appropriate extension of the model is given by:

$$\boldsymbol{Bias}(y_r) = \left(\frac{rf}{n}\right)E(y_r - y_{rf}) + \left(\frac{nl}{n}\right)E(y_r - y_{nl}) + \left(\frac{o}{n}\right)E(y_r - y_o)\,, \quad (2.2)$$

where

y_r = estimate based on the r respondent cases;
y_{rf} = estimate based on the refusal cases;
y_{nl} = estimate based on the nonlocated cases;
y_o = estimate based on the other nonresponse cases;
n = total sample size;
rf = number of refusals
nl = number of nonlocated cases;
o = number of other nonresponse cases.

The three major components of nonresponse in the RCG:91, as shown in (2.2) are graduate refusals, being unable to locate the graduates for an interview, and all other nonresponse. The nonlocatables account for 62 percent of the overall nonresponse rate and the refusals account for another 30 percent, while the other category is only 8 percent of the nonresponse.

Different estimates of the nonresponse bias can be computed using (2.2), based on data available on the frame. These estimates were not calculated for the RCG:91 because the sample size was too small for all but the nonlocatable category, and the resulting bias estimates would be subject to large sampling errors.

Nonresponse Bias after Adjustments

So far, the two models presented assumed that the estimates were not adjusted to reduce the level of bias arising from graduate nonresponse. In the RCG:91, graduate nonresponse adjustments and poststratification adjustments were used specifically for this purpose. These adjustments resulted in estimates of characteristics that are different from the simple estimates suggested by (2.1) and (2.2).

The weighting adjustment process is depicted below.

$$y_r \rightarrow\rightarrow\rightarrow\rightarrow y_r^* \rightarrow\rightarrow\rightarrow\rightarrow y_r^{**}$$

where y_r is the estimate adjusted at the institution level only (includes components a-d described on page 2-8), y_r^* is the estimate adjusted for graduate level nonresponse (components a-e), and y_r^{**} is the estimate adjusted for both graduate-level nonresponse and for poststratification (components a-f).

The adjustment for graduate-level nonresponse (component e) was done by forming nonresponse adjustment cells based on known characteristics of the sampled graduates (degree, race/ethnicity, and major). Note that these characteristics were those reported by the institutions for all bachelor's degree recipients and did not necessarily match data reported by the respondent on the survey. In each cell, the ratio of the number of sampled graduates to the number of responding graduates was used to adjust the weight for all graduates in the cell. This can be written as:

$$y_r^* = \sum_{i=1}^{r} \sum_{k=1}^{c} \frac{n_k}{r_k} y_{ik} \qquad (2.3)$$

where n_k is the number of sampled graduates in adjustment cell k, r_k is the number of respondents in adjustment cell k, and y_{ik} is the characteristic of the ith responding graduate in adjustment cell k.

In essence, the adjustment is equivalent to estimating the bias as in (2.1) and adjusting the weights to remove this bias. In fact, the graduate-level nonresponse adjustment eliminates the bias in the estimated number of graduates with the characteristic if the adjustment cell is identical to the characteristic. Going back to model (2.1), we can show this by writing:

$$Bias(y_{r,k}) = \left(\frac{n-r}{n}\right) E(y_{r,k} - y_{nr,k}) \doteq \left(\frac{n-r}{n}\right)(r_k - n_k)$$

$$Bias(Y^*_{r,k}) = \left(\frac{n-r}{n}\right) E(y^*_{r,k} - y_{nr,k}) \doteq \left(\frac{n-r}{n}\right) \left(\frac{n_k}{r_k} r_k - n_k\right) = \mathbf{0} \quad (2.4)$$

where $y_{r,k}$ = estimate of the number of units in adjustment cell k based on the r respondent cases.

This formulation demonstrates that the bias is removed completely for the characteristics used to define the nonresponse adjustment cells. For characteristics correlated with the variables used to define the cells, the nonresponse bias is generally attenuated. An analogy is the reduction in variance due to the use of independent variables in a regression problem, where in this case the bias is reduced due to the introduction of the nonresponse adjustment variables.

In addition to the nonresponse adjustment, the estimates in the RCG:91 were poststratified by gender, major, degree, and institutional control using counts from IPEDS.[3] This estimation procedure resulted in estimates that are equal to the control totals for these variables. Thus, these estimates from the RCG:91 are fixed and have no sampling error.

Poststratified estimates enable us to look at the unconditional bias of the estimate. Recall that in expression (2.1) the bias was written as an expectation over all possible samples. When we computed estimates of the bias from the sample data, we evaluated y_r and y_{nr} for the specific sample observed, i.e., conditional on the sample of institutions and graduates for the RCG:91. Another way of writing (2.1) is:

$$Bias(y_r) = E(y_r) - E(y_n) \quad , \qquad (2.5)$$

where y_n is the estimate based on all the sampled cases.

We can replace the expectation of the estimate for all sampled units by the known population totals for the poststratification variables. In other words, the $E(y_n)$ in expression (2.5) can be replaced by the known population total for the poststratification variable. This value is not conditioned on the specific sample selected for the RCG:91.

The extension of (2.5) to account for poststratification is direct. The difference between the nonresponse adjusted estimate and the poststratified

[3]U.S. Department of Education, National Center for Education Statistics, Integrated Postsecondary Education Data System (IPEDS) "Completions" Survey, 1989-90.

estimate (y_r^*-y_r^{**}) for the variables used in poststratification provides an estimate of the unconditional size of all the errors due to sources other than nonresponse (e.g., sampling, noncoverage, and measurement error). This follows because y_r^* is already adjusted for nonresponse for these variables and the poststratification adjustment is only correcting for other sources of error.

Comparing the relative size of the difference between the estimates (y_r-y_r^* and y_r^*-y_r^{**}) provides an indication of the potential for nonresponse bias from the RCG:91. If the difference between the nonresponse adjusted estimate and the poststratified estimate (y_r^*-y_r^{**}) is large relative to the other difference (y_r-y_r^*), it indicates that sampling and measurement errors are more important problems than the bias due to nonresponse. On the other hand, if the difference between the poststratified and the nonresponse-adjusted estimates (y_r^*-y_r^{**}) are relatively small, then the nonresponse bias should be considered a potentially major source of error.

Below, the data available from the RCG:91 are used to estimate the relative sizes of the errors and to indicate the potential for bias due to nonresponse, using the models and statistics described above.

ESTIMATES OF GRADUATE NONRESPONSE BIAS

The first application of these methods is the modeling of the nonresponse bias using expression (2.1). Table 2-4 shows the estimates of the bias, the relative bias, and the bias ratio for the variables for which these estimates could be computed. As mentioned before, these estimates rely on data for both the respondents and the nonrespondents and could only be calculated for the items that were available on the frame.

The first column of the table shows the estimated percentage of graduates in each category, based only on the respondent data before nonresponse adjustment at the graduate level (y_r). This is provided for reference purposes. The other quantities in the table were computed as discussed in the previous section. The standard errors of the estimates used in the bias ratio were computed based on the sample design.

The estimated biases and relative biases of the estimates are generally relatively small. The only variable where the relative bias exceeds 5 percent is race/ethnicity. For this variable, the relative biases for the Asian and black subgroups are 10 percent and 18 percent, respectively. The high relative bias in the estimates for Asian and black graduates is due to the combination of the higher nonresponse rate in these subgroups than in the overall population and the differences in the estimated percentages between respondents and nonrespondents. The Asian and black graduates are the only subgroups with response rates less than 80 percent (see Table 2-2).

For most of the variables of interest, the estimated bias ratios are relatively high. For the variables with high ratios, the impact due to the bias is large primarily because the RCG has very large sample sizes for estimates of aggregates. These large sample sizes yield estimates with small standard errors. Nonresponse bias can dominate sampling errors in the RCG for many

Table 2-4. Estimated bias and relative bias in the RCG from graduate nonresponse, by graduate characteristic

Graduate characteristic	Estimated percent based on respondents[1]	Estimated bias	Estimated relative bias	Ratio of bias to standard error
Degree				
Bachelor's	78.0%	0.3%	0.4%	84.4%
Master's	22.0	-0.3	-1.5	-84.4
Major*				
Education	10.5	0.5	4.3	147.8
Math	1.6	0.1	4.8	61.2
Physical Sciences	1.6	0.0	2.4	29.9
Other	86.3	-0.6	-0.7	-165.1
Institution control				
Public	66.9	-0.3	-0.5	-73.2
Private	33.1	0.3	1.0	73.2
Gender				
Male	45.9	-0.3	-0.6	-57.7
Female	54.1	0.3	0.5	57.7
Race/ethnicity[2]				
Native American	0.3	0.0	0.3	1.5
Asian/Pacific Islander	2.6	-0.3	-10.3	-175.5
Black, non-Hispanic	3.8	-0.7	-18.0	-175.7
Hispanic	1.8	-0.1	-3.7	-31.3
White, non-Hispanic	62.0	2.2	3.6	440.7
Not reported	29.6	-1.2	-4.2	-272.0
Institution size class				
<1,500	38.8	0.6	1.7	140.1
1,500-5,999	49.3	-0.5	-1.0	-105.9
6,000+	12.0	-0.1	-1.2	-47.3

[1]This estimate does not include any adjustments for graduate nonresponse or poststratification. It does contain the school nonresponse adjustment.

[2]Major and race/ethnicity were not collected for master's degree graduates.

NOTE: Percentages may not add to 100 due to rounding.

SOURCE: U.S. Department of Education, National Center for Education Statistics, *1991 Recent College Graduates Survey.*

aggregate statistics based on all graduates. For estimates of smaller subgroups, like the percentage of mathematics or physical sciences majors, the sampling errors are larger and the bias ratios smaller, indicating that nonresponse has less impact on the error of these types of estimates.

These findings tentatively indicate that the nonresponse bias could be a major problem in the study and might argue for allocating more resources for the reduction of nonresponse, even at the expense of decreasing the sample size. However, before concluding this, we need to examine the impact of the nonresponse adjustments. Before doing this, we briefly present some estimates related to the reasons for nonresponse.

The expanded model for nonresponse bias (2.2) incorporated different reasons for graduates not completing the interview. As we mentioned before, different estimates of bias based on model (2.2) would be subject to substantial sampling errors and are not presented. However, it is instructive to examine the components of that model (the percent of nonresponse by reason and the difference in the estimates based on the respondents and each group of nonrespondents).

Table 2-3 gives the nonresponse rate (100 minus the response rate) and the percent distribution of nonresponse by the three major reasons. Even though the distributions of the total nonresponse by reason are of the same magnitude from one variable to the next, there is important variability that could increase the bias due to nonresponse. For example, the percent of nonresponse due to not being able to locate black (78 percent) and Hispanic (82 percent) graduates is large compared to the 62 percent for all graduates (Figure 2-5).

Figure 2-5. Percentage distribution of type of nonresponse by race/ ethnicity: 1991, RCG

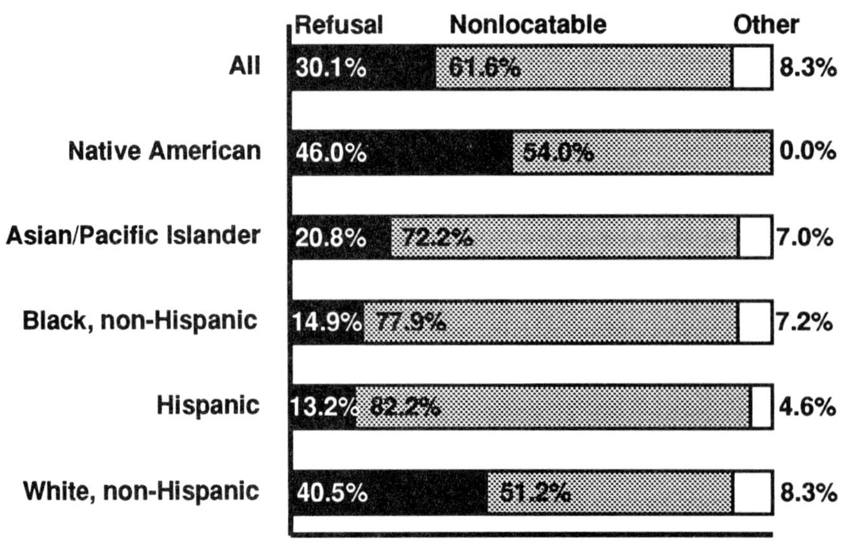

Table 2-5 completes this examination of reasons for nonresponse bias by presenting the differences between the estimates based on the respondent sample and the nonrespondents for various groups. The first column of the table gives the estimate based on the respondents to provide a benchmark for assessing the size of the differences. As before, the differences are relatively small except for the estimates by race/ethnicity.

Estimated Bias after Adjustments

Since the estimates are adjusted for nonresponse and poststratified, a critical part of the analysis is based on the adjusted estimates. Table 2-6 presents the estimates before graduate nonresponse adjustment (y_r), after graduate nonresponse adjustment (y_r^{*}), and after poststratification (y_r^{**}).[4] The differences between the estimates (y_r-y_r^{*}) and (y_r^{*}-y_r^{**}) are given in the last two columns of the table.

As noted in the last section, if the difference in the last column is small relative to the difference in the next to last column, then nonresponse bias could be considered a potentially major source of error relative to other errors in the survey. Conversely, when the difference in the last column is large relative to that in the next to last column, the nonresponse bias may not be as important as other sources of error.

This interpretation of the difference is technically valid when the estimate is the aggregate of the number of graduates in a cell used for nonresponse adjustment. Since degree, major and race/ethnicity were the only three items used for defining nonresponse adjustment cells, these are the only ones that will be examined from this perspective. The estimated bias shown in Table 2-4 is nearly equal to the difference y_r-y_r^{*} shown in Table 2-6 for these variables, as would be expected since they were used in the nonresponse adjustment.

For race/ethnicity, the estimate in the next to last column, (y_r-y_r^{*}), for blacks is -0.7, while the difference in the last column, (y_r^{*}-y_r^{**}), is less than 0.05. This finding indicates that the potential for nonresponse bias is substantial for this estimate. For the other two variables, major and degree, there is less potential for substantial nonresponse bias, since the estimated differences in the last column are of the same size or larger than those in the next to last column.

[4]The poststratified estimates shown in Table 2-6 are different from the usual RCG estimates because the sample characteristics of the graduates were taken from the sampling lists for these tabulations and were then poststratified to IPEDS totals. The graduate-reported characteristics are used in reports and all other tabulations. The data from the sampling lists had to be used for this assessment because graduate reports were not available for the nonresponding graduates.

Table 2-5. Differences in the estimates between respondents and nonrespondents, by type of nonrespondent and characteristic of graduate

Graduate characteristic	Estimate based on respondents	Difference between estimates based on respondents and nonrespondents		
		Refusals	Nonlocatables	Other nonrespondents
Degree				
Bachelor's	78.0%	-0.6%	3.3%	1.4%
Master's	22.0	0.6	-3.3	-1.4
Major*				
Education	10.5	2.6	2.9	2.6
Math	1.6	0.4	0.5	0.2
Physical Sciences	1.6	0.8	-0.1	0.7
Other	86.3	-3.7	-3.3	-3.5
Institution control				
Public	66.9	2.9	-4.6	0.1
Private	33.1	-2.9	4.6	-0.1
Gender				
Male	45.9	1.1	-3.1	-0.1
Female	54.1	-1.1	3.1	0.1
Institution size class				
<1,500	38.8	4.9	3.3	3.7
1,500-5,999	49.3	-4.7	-2.5	-0.2
6,000+	12.0	-0.3	-0.8	-3.5
Race/ethnicity*				
Native American	0.3	-0.1	0.0	0.3
Asian/Pacific Islander	2.6	-0.2	-2.4	-0.9
Black, non-Hispanic	3.8	0.0	-6.5	-3.1
Hispanic	1.8	0.9	-1.2	0.6
White, non-Hispanic	62.0	-0.8	21.1	14.0
Not reported	29.6	0.2	-11.1	-10.8

*Major and race/ethnicity were not collected for master's degree graduates.

NOTE: Percentages may not add to 100 due to rounding.

SOURCE: U.S. Department of Education, National Center for Education Statistics, *1991 Recent College Graduates Survey.*

Table 2-6. Graduate nonresponse adjusted and poststratified estimates, by graduate characteristics

Graduate characteristic	Estimate based on respondents[1] (y_r)	Estimated after graduate nonresponse adjustment (y_r*)	Poststratified estimate (y_r**)	Difference between estimates	
				y_r-y_r*	y_r*-y_r**
Degree					
Bachelor's	78.0%	77.8%	76.4%	0.2%	1.4%
Master's	22.0	22.2	23.6	-0.2	-1.4
Major[2]					
Education	10.5	10.2	10.3	0.3	-0.1
Math	1.6	1.5	1.4	0.1	0.1
Physical Sciences	1.6	1.5	1.3	0.1	0.2
Other	86.3	86.8	87.0	-0.5	-0.2
Institution control					
Public	66.9	66.8	64.4	0.1	2.4
Private	33.1	33.2	35.6	-1.1	-2.4
Gender					
Male	45.9	45.9	47.0	0.0	-1.1
Female	54.1	54.1	53.0	0.0	1.1
Institution size class					
<1,500	38.8	38.8	39.6	0.0	-0.8
1,500-5,999	49.3	49.2	48.8	0.1	0.4
6,000+	12.0	12.0	11.6	0.0	0.4
Race/ethnicity[2]					
Native American	0.3	0.3	0.3	0.0	0.0
Asian/Pacific Islander	2.6	2.6	2.5	0.0	0.1
Black	3.8	4.5	4.5	-0.7	0.0
Hispanic	1.8	1.8	1.8	0.0	0.0
White	62.0	61.5	61.3	0.5	0.2
Not reported	29.6	29.4	29.6	0.2	-0.2

[1]This estimate does not include any adjustments for graduate nonresponse or poststratification. It does contain the school nonresponse adjustment.

[2]Major and race/ethnicity were not collected for master's degree graduates.

NOTE: Percentages may not add to 100 due to rounding.

SOURCE: U.S. Department of Education, National Center for Education Statistics, *1991 Recent College Graduates Survey.*

The other three variables in Table 2-6 were not used in the formation of the nonresponse adjustments and the relative magnitudes of the estimates in the last two columns cannot be interpreted in the same fashion. For these variables, the estimated bias shown in Table 2-4 is generally larger than the estimated difference $(y_r - y_r^*)$ given in Table 2-6. This finding suggests that the bias in these variables is not fully accounted for by the nonresponse adjustment cells.

The differences in the last column of Table 2-6 are generally greater than the differences in the next to last column for the three variables (gender, institutional control, and institution size) that were not used in forming the nonresponse adjustment cells. The evidence is very limited, but this might happen because the poststratification is handling both nonresponse bias and other types of errors for these variables.

IMPLICATIONS AND RECOMMENDATIONS ABOUT GRADUATE NONRESPONSE BIAS

The findings presented above are somewhat limited because they are based on so few variables and the variables are ones that may not be particularly impacted by differential nonresponse. For example, there may be serious nonresponse bias for particular items that are uncorrelated with the variables used in the nonresponse and poststratification adjustments. It is not possible to examine these types of issues in more depth due to the lack of data about nonrespondents.

Despite these limitations, some general comments are possible. The simple estimates of nonresponse bias before adjustments show that there is a significant potential for nonresponse bias that could significantly reduce the nominal level of confidence intervals or statistical tests. These types of nonresponse bias are more likely to occur when the results are for estimates based on large sample sizes, because the sampling errors are smaller for these estimates. For estimates of smaller subdomains, the impact of nonresponse bias may be less important, provided the nonresponse is not correlated with membership in the subdomain. As discussed below, race/ethnicity does not fall into this category.

This general finding is likely to hold for many statistics, including those that could not be investigated in this study. Data users should be particularly aware of the potential for nonresponse bias if there is a correlation between response rates and having the characteristic. In other words, if some evidence or theory implies that the response rate is likely to be much higher (or lower) than average for the persons with the characteristic (e.g., being a teacher), it is possible that the nonresponse bias could be substantial for estimating this characteristic.

The results from the study of the adjusted estimates show the adjustments may substantially reduce the nonresponse bias in the estimates. The nonresponse bias does not appear to be a dominant source of error after the adjustment, although it is clearly still important. Particular problems were noted for the race/ethnicity estimates.

Data users should be generally encouraged by these findings. Many of the specific variables estimated by users will be correlated with the variables used in the nonresponse adjustments and poststratification adjustments. The adjustments should provide good protection for most of these items, since the residual nonresponse bias is likely to be of a lower magnitude. The main concern for data users involves estimates that are not correlated with the variables used in the adjustments, especially those that are associated with differential response rates. If users suspect this condition holds and that the nonresponse is correlated with having a specific characteristic, it may be wise to employ conservative inference procedures (e.g., use 99 percent confidence intervals rather than 95 percent ones).

For more analytic estimates (regressions and correlations) of characteristics that meet these conditions, users may wish to include explanatory or control variables that mediate the impact of nonresponse bias. For example, including variables that are thought to be correlated with the dependent variable and with the response rate might be useful for regression analyses.

Some recommendations for future studies can also be drawn from these findings. Efforts to reduce nonresponse bias in the future surveys should concentrate on those groups of nonrespondents that both have larger than average nonresponse rates and exhibit large differences in characteristics between respondents and nonrespondents. The Asian and black graduates satisfy these conditions for the current RCG sample. Furthermore, since the nonlocatables account for most of the graduate nonresponse, the findings imply that it would be worthwhile to consider investing more resources in tracing elusive groups of graduates. For Asian and black graduates, locating was a particular problem.

Another possibility to deal with the estimates by race and ethnicity is to consider poststratifying the estimates from the RCG to the IPEDS totals for these categories. However, this is not recommended without further study and evaluation. The error characteristics, including the completeness of the reporting for these items, of the IPEDS for estimates by race and ethnicity need to be considered. The summary comparison of the RCG:91 and the IPEDS estimates[5] in Table 2-7 shows that the RCG estimate of Hispanic graduates is greater than the IPEDS figure, and the difference is statistically significant. If the IPEDS totals are biased downward for these characteristics due to imputation for missing data or any other reasons, then this bias would be transferred to the RCG estimates if race and ethnicity variables were used in poststratification.

Another recommendation is to consider a special study of nonrespondents to evaluate a number of characteristics that were not available in this assessment, especially key characteristics such as being newly qualified to teach or

[5]The estimated percentage was computed excluding nonresident aliens and those with unknown race/ethnicity in IPEDS for 1989-90.

Table 2-7. Percent of bachelor's recipients from 1989-90 IPEDS completions file, and RCG estimates with standard errors of bachelor's recipients, by race/ethnicity

Race/ ethnicity	IPEDS totals*	RCG	
		Estimates	Standard error
Black	6.0%	6.1%	0.4%
Hispanic	3.2	3.8	0.2
Other	90.8	90.1	0.5

*Includes continental United States only.

SOURCE: U.S. Department of Education, National Center for Education Statistics, *1991 Recent College Graduates Survey.*

being unemployed. A more intensive effort to locate and interview a subsample of nonrespondents could provide evidence of patterns of nonresponse bias for characteristics beyond that available from the frame.

Of course, this type of special study is not without its problems. The main problem with an intensive study of nonrespondents is the expected response rate. The 30 percent of the nonrespondents who refused may still refuse the followup. Furthermore, over 60 percent of the nonresponse in the RCG:91 was due to not being able to locate the graduate for an interview. These hard-to-locate cases are expensive to complete and this might limit the sample size that can be included in the intensive study. If the sample size or the response rate for the followup is too low, the results may be less than conclusive.

Another approach to this idea is to collect more data on the sampled graduates from the institutions. In fact, transcripts of the sampled graduates (both respondents and nonrespondents) were collected. These data could be used to examine the nonresponse bias in greater detail. However, many key characteristics, such as labor force status and being newly qualified to teach, cannot be obtained from transcripts.

ITEM RESPONSE RATES AND IMPUTATION

In addition to unit nonresponse, bias may result from item nonresponse. Item nonresponse is the failure to obtain a valid response for a particular item even though the graduate completed most of the items in the survey. Item nonresponse may occur because a respondent does not wish to answer a specific item or when the response obtained is later found to conflict with the responses given to other items in the interview.

Since the RCG:91 was a telephone survey using CATI methods, the item response rates were typically very high. The item response rates for nearly all of the items were greater than 98 percent. They are shown in Table 2-8. These high item response rates are typical for well-designed telephone surveys that include online checks for both range errors and logical inconsistencies.

The only items with relatively low item response rates are ones that are explained in the footnotes of the table. Virtually all of these lower item response rates are for items that were asked only of a small portion of all graduates, due to skip patterns. Even a small amount of incomplete data could result in a low item response rate in these cases.

Whenever data were missing, the values were imputed. The imputations were done to make analysis simpler and more consistent. Item imputation may also reduce the nonresponse bias. The first step of imputation was to determine if any of the missing values could be inferred from the responses to other items in the interview. This type of logical imputation was limited to a very few items.

Most of the missing values were replaced by a hot-deck imputation procedure. The interviews that contained a valid response to the item were sorted by other variables thought to be correlated to the missing item. Within the subgroups formed by these sort variables, an interview was selected and the response from the selected interview was "donated" (assigned to replace the missing valuc). This process was repeated for each record with a missing value, with controls to prevent the same interview from being selected as a donor too often. The hot-deck imputation process was done for each variable with missing values on the file. The sort variables were specified uniquely for each item to be imputed.

As a result of the imputation, none of the items on the file have missing values. Each imputed value is identified by a flag that indicates it was imputed and the type of imputation that was done.

Implications and Recommendations About Item Nonresponse

Item nonresponse can be modeled in much the same way as for unit nonresponse. The simple model for unit nonresponse bias (2.1) could be applied to item nonresponse, with the unit response rate replaced by the item response rate. Since the item response rates are so high for the RCG:91 (almost all in excess of 98 percent), this model shows that the possibility of substantial bias from item nonresponse is very small.

The imputation for missing values is also equivalent in some sense to an adjustment for nonresponse. Therefore, the size of the item nonresponse bias estimated by a model such as (2.1) is larger than would be obtained using the imputed values. Just as for unit nonresponse, the adjustment for missing values (the imputed values) should reduce the bias in a manner similar to that presented in (2.5). This reduction in the bias occurs if the variables used in the imputation process were correlated with the missing item. As discussed above, the variables were chosen specifically with this goal in mind.

Table 2-8. Weighted item response rates

Question	Description	Number of Eligibles	Weighted Item Response Rate
Q1	Name confirm	14,405	100.0
Q2	School confirm	14,405	100.0
Q3	Date confirm	14,405	100.0
Q4	Was this degree bachelor's or master's?	14,405	100.0
Q5A	Date when degree was received	14,405	100.0
Q6	Major field of study	14,405	100.0
Q7A	Minor field	12,888	99.9
Q7B	Minor field of study	4,426	100.0
Q8A	Second major field	12,888	100.0
Q8B	What was second major?	1,164	100.0
Q9A	Undergraduate major field	1,517	99.8
Q9B	Was there an undergraduate minor?	1,517	99.8
Q9C	Undergraduate minor field of study	574	100.0
Q9D	Undergraduate second major field?	1,517	99.9
Q9E	What was undergrad second major?	200	100.0
Q10	Gradepoint average for undergrad level	14,405	99.0
Q11	Did R apply for additional training?	14,405	100.0
Q12	Has R attended since receiving degree?	14,405	100.0
Q13	Best reason for not applying for training	8,337	99.4
Q14	Date first attended	5,032	99.4
Q15	Is respondent still enrolled?	5,032	100.0
Q16	Date R stopped attending	1,903	99.5
Q17	Type of school R was attending	5,032	99.4
Q18	Is this a public or private institution?	5,032	99.2
Q19	What degree was R working toward?	5,032	100.0
Q20	Date for obtaining degree	4,067	91.9
Q21	Major field of study for further degree	5,032	99.4
Q22	Was R attending full or part time?	5,032	99.9
Q23A	Did R have assistantships or CWS?	5,032	98.8
Q23	Was R working for pay in reference week?	14,405	99.8
Q24	Was R looking for work in reference week?	2,165	99.4
Q25	Was R available to work in reference week?	2,165	99.1
Q26	What was main reason for not working?	2,165	99.4
AQ27	Industry verbatim	12,240	99.9
AQ28	Occupation verbatim	12,240	99.9
Q28VERF	Job verification of Q28	12,240	99.9
AQ29	Duties on job	12,240	99.8
Q31	Miles from home when senior	12,240	99.5
Q32	Was this job full time or part time?	12,240	99.8
Q33	Would R have wanted full-time job?	1,633	96.6
Q34	What kind of employee was respondent?	12,240	99.0
Q35	Was business incorporated or not?	322	98.2

Table 2-8. Weighted item response rates--(continued)

Question	Description	Number of Eligibles	Weighted Item Response Rate
Q36	Hours/week respondent worked in business	322	97.6
Q37	Annual income from business before taxes	322	84.1*
Q38	Hours/week R usually employed at job	11,918	99.4
Q39	Income from principal job	11,918	94.6
Q40	Was R working for pay at second job?	12,240	99.5
Q41	Was second job as school teacher?	1,581	99.4
Q42	Was college degree required for main job?	12,240	99.0
Q43	How close was major related to main job?	12,240	99.7
Q44	Main reason job not related to major	2,629	97.9
Q45	What best describes job/career on Apr 22?	12,240	99.6
Q46	Was R looking for another job -Apr 22?	12,240	99.7
Q47A	Was there work experience before degree?	14,405	100.0
Q47B	Was work experience full or part time?	12,786	99.6
Q48	Full-time work permanent or summer job	8,496	99.5
Q49	Experience in permanent jobs before degree	5,008	99.4
Q50	Is R eligible to teach at any level?	14,405	99.4
Q51	Grade(s) R is eligible to teach	3,238	100.0
Q52	When did R first become eligible?	3,238	99.7
Q53	Does R have certificate to teach school?	14,405	99.9
Q54	Grade(s) R has certificate for	3,111	100.0
Q55	Date R got certificate to teach	3,111	97.2
Q56	Kind of certificate or license R has	3,111	99.2
Q57A	Is certification issued by state?	3,111	99.8
AQ57B	Teacher certification agency - State	3,086	99.7
AQ57CANC	Teacher certification agency - Name	25	100.0
AQ57CAST	State of agency certification	25	100.0
Q58	Field(s) eligible to teach	454-3,238	97.7-99.9
Q59	Field(s) certified to teach	350-3,111	96.6-99.6
Q60	Which field is R best qualified in?	2,910	98.2
Q61	Has R ever taught any grade?	14,405	99.7
Q62	Before degree, was R employed as teacher	14,405	99.7
Q63	Was R employed as teacher full/part time?	1,028	100.0
Q64	Has R ever applied for job as teacher?	14,405	100.0
Q65	Main reason R did not apply for teacher?	11,316	99.1

*Annual income from business (Q37) was asked only of graduates who were self-employed. Since only 2.5 percent of the graduates were self-employed, the number of graduates for whom this question was applicable was small. This question was used in conjunction with Q39 (salary rates for all other graduates) and Q87C (annual income for teachers under contract) in all published reports. The overall response rate for salary for all working graduates was 94.3 percent.

NOTE: Item response rates were calculated as the weighted number of respondents who answered a given item divided by the weighted number of respondents for whom the item was applicable.

Table 2-8. Weighted item response rates--(continued)

Question	Description	Number of Eligibles	Weighted Item Response Rate
Q66	Has R taught any grade since degree?	14,405	99.2
Q67	Date when R started teaching	2,988	98.9
Q68	Principal job as school teacher, any level	14,405	99.3
Q69	Which grade(s) did R teach?	2,330	98.7-100.0
Q70	Types of schools R taught in	2,330	99.8
Q71	Field(s) R was teaching in	2,330	99.2-99.3
Q72	What field did R teach most of the time?	1,249	89.1
Q73	Any fields not adequately prepared in	2,327	99.1
Q74A	Which field(s) not prepared to teach?	339	97.3
Q74B	Teach in self-contained classroom	2,327	98.7
Q75A	Has R received training - Bilingual Ed?	2,330	99.9
Q75B	Has R received training - ESL?	2,330	99.8
Q75C	Has R received training - LEP?	2,330	99.9
Q76	Has R taught students in LEP?	2,330	99.7
Q77	Number of LEP students taught	760	96.9
Q78A	Has R taught classes - Bilingual Ed?	760	99.3
Q78B	Has R taught classes - ESL?	760	99.5
Q78C	Has R taught classes - LEP?	760	99.5
Q79	How well R prepared to teach LEP classes?	760	99.1
Q80	Did R teach Special Ed students?	2,330	99.8
Q81	Did R teach primarily in Spec Ed?	1,814	99.7
Q82	Was R teaching other than Spec Educ?	286	100.0
Q83	Did R take Spec Ed courses for credit?	1,814	99.8
Q84A	Did R have training in Spec Ed?	1,814	99.8
Q84B	Did R feel prepared to teach Spec Ed?	1,814	99.6
Q85	Was teaching assignment full/part time?	2,330	99.9
Q86	What level was part-time teaching assign?	363	97.8
Q87	Have teaching contract/other arrange?	2,330	92.8
Q87A	# of mths per year was teaching contract	1,803	99.3
Q87B	# of months paid for teaching	1,803	99.5
Q87C	Annual teaching income	1,803	94.9
Q87D	Any summer employment besides teaching	1,803	99.3
Q87E	Income from summer employment	658	96.6
Q88	Reason for R becoming teacher	2,330	95.0
Q89A	Does R expect to teach 1991-92 year?	2,330	99.3
Q89B	Primary reason for not teaching next yr	225	100.0
Q90	Date of birth of respondent	14,405	99.5
Q91	Gender of respondent	14,405	100.0
Q92	Is R a U.S. citizen?	14,405	100.0
Q93	Is R a resident?	395	98.9
Q94	Is R of Hispanic origin?	14,405	99.9
Q95	What is R's race?	14,405	98.1

Table 2-8. Weighted item response rates--(continued)

Question	Description	Number of Eligibles	Weighted Item Response Rate
Q96	What was R's marital status in April 91?	14,405	99.8
Q97	How many dependent children does R have?	14,405	99.9
Q98	Did R receive HS diploma, GED, or other?	14,405	99.9
Q99	What year did R receive diploma?	14,405	99.7
Q100	Year began working towards bachelor's	14,405	99.8
Q101	Highest level of education expected	14,405	97.9
Q102	Highest grade R's father completed	14,405	99.0
Q1024	Vocational school R's father completed	294	91.5
Q1025	College education R's father completed	7,699	98.6
Q103	Q103 father/male guardian occupation	14,405	98.1
Q104	Highest grade R's mother completed	14,405	99.1
Q1044	Vocational education R's mom completed	708	93.8
Q1045	College education R's mom completed	6,952	98.8
Q105	Q105 mother/female guardian occupation	14,405	99.2
Q106	Expenses paid by	14,405	99.6-99.7
Q107	Percent paid by	51-10,933	94.9-99.3
Q108	Did R ever apply for financial aid?	14,405	99.6
Q109A	Was work study used to finance degree?	2,920	98.9
Q109B	Was fellowship used to finance degree?	260	95.9
Q109C	Was assistantship used to finance degree?	496	98.7
Q110	Types of grants/scholarships received	1,221-7,785	77.5**-97.3
Q111	Types of loans received	455-6,411	94.9-99.0
Q112	Total amount of federal money borrowed	5,619	94.6
Q113	Has R consolidated loans?	5,619	97.9
Q114A	Monthly payments for GSL loan	3,196	93.8
Q114B	Consolidated monthly loan payment	2,423	95.2
Q115	Total amount borrowed by respondent	6,889	94.2
Q116	Total amount owed by respondent	6,889	91.7
Q117	When will R repay all loans - year?	6,300	86.4

SOURCE: U.S. Department of Education, National Center for Education Statistics, *1991 Recent College Graduates Survey.*

** The item response rate of 77.5 occurred in item Q110B2 (Did R receive other federal grants or scholarships between July 1, 1989 and June 30, 1990). There were 252 (unweighted) cases imputed for this item. Of these, 139 were imputed because the information reported in Q110 (type of grants or scholarships) was not consistent with Q106H (grants/scholarships from federal, state or local government, or college or university). Specifically, if the graduate reported having a grant/scholarship in Q106, but reported "no" to every category in Q110, then every category in Q110 was changed to "not ascertained" and later imputed by hot deck. While the number of cases imputed in Q110B2 is about the same as the number imputed in the rest of Q110, the number of cases for which Q110B2 is applicable (1,221) is much smaller. Therefore, the response rate for Q110B2 is much lower than for the rest of Q110. The item response rates for the rest of Q110 range from 92.8 to 97.3.

The implications for data users for item nonresponse are relatively minimal. For most of the items, item nonresponse will result in trivial bias. Special concerns do exist for the few items that have item response rates of less than 95 percent. These are identified in Table 2-8. In addition, users doing multivariate analysis may wish to evaluate the number of items that have missing values to ensure that the potential for item nonresponse bias is small. Even in these situations, the use of the imputed values should make this type of analysis more complete and less subject to bias.

A typical problem associated with using imputed values that data users face is in the estimation of the precision of the estimates. Imputed values are often treated as if they were valid responses in estimating the standard errors of the estimates. This tends to inflate the estimates of the precision of the estimates. However, for the RCG:91 the high item response rates effectively eliminate this problem also.

The use of the CATI in the RCG:91 resulted in very high item response rates across a broad spectrum of the variables. For the few items with lower item response rates, specific edit and consistency checks can be added to improve the responses. Overall, the basic procedures used in the RCG:91 were very effective in reducing the bias due to item nonresponse.

3. NONSAMPLING ERROR FROM MEASUREMENT ERROR: REINTERVIEW MEASURES

Measurement problems are an inevitable source of errors in any survey or census. Measurement errors, sometimes called response errors, depend upon a number of factors, including imperfect instructions to the interviewers, unclear questions, respondent recall problems, coincidental factors that affect the interviewer and/or respondent during the interview, or deliberate errors by the respondent or interviewer.

In addition to arising from a wide variety of sources, measurement errors are difficult to estimate. One way of estimating the size of measurement errors is by using an external source of data to validate the findings. This approach is discussed in Chapter 5 for the certification to teach variable.

Another way to estimate measurement errors is by interviewing the respondents again. A special reinterview study was conducted with the RCG:91 in an attempt to estimate the impact of measurement errors on estimates from the survey. The reinterview program for the RCG:91 entailed calling a sample of respondents who had previously completed the RCG:91 interview and asking them a portion of the interview questions again. At the end of the reinterview, graduates were asked to reconcile differences between the original and reinterview responses for a subset of the reinterview questions. Items chosen for reconciliation were those considered key items, typically those that have been used in reports in the RCG series.

In this chapter, we present models for estimating measurement errors from the reinterview data. We begin by using a model that assumes the errors are all from random sources. Estimators of the parameters of this model are then presented. The model is then expanded to allow for systematic errors or biases. Estimators for this model are also given. The two models are then applied to the RCG:91.

The two models presented assume that the interviewers are not a source of systematic error in the data collection process. In the next chapter, we relax this condition and assess the contribution of interviewers to measurement errors.

PURPOSE OF THE REINTERVIEW

Reinterview programs have been employed in other surveys to detect falsification by interviewers, to evaluate field work, and to estimate response bias and variance. Response bias and variance are technical terms that refer, respectively, to systematic errors and random errors and are defined later in this chapter. Since the RCG:91 was done using CATI in a centralized setting, the reinterviews were not needed to verify that the interviews were genuine. The CATI interviews were closely monitored and it was highly unlikely that a telephone interviewer could invent or falsify interviews.

The primary objectives for the RCG:91 reinterview study were concerned with estimating response bias and variance. Specifically, the goals were

- To identify survey items that were not reliable;
- To quantify the magnitude of the measurement error; and
- To provide feedback on the design of questionnaire items for future surveys.

Forsman and Schreiner (1991) show that the optimal design for a reinterview study depends on the purpose of the reinterview. The design of the RCG:91 reinterview attempted to maximize the ability to estimate the random component of the measurement error within the context of a limited study.

REINTERVIEW DESIGN

The RCG:91 reinterview was a one-stage sample of the original interview respondents and had a goal of 500 completed reinterviews. The respondent for the reinterview was the same graduate as in the original survey in all cases. A simple random sample of 583 respondents was selected from the eligible RCG:91 survey completions. Only graduates that met the following criteria were eligible for the reinterview sample:

- Bachelor's degree recipients;
- Graduates who had never refused to participate; and
- Graduates who were interviewed for the main survey between August 15 and September 30. The reason for this eligibility time period was to exclude respondents interviewed in the first 3 weeks of data collection (during which time interviewers were learning the survey), and to establish a cutoff for data collection to ensure that at least 2 weeks had elapsed between the first and second interviews.

Of the 583 respondents sampled for the reinterview, 512 completed the reinterview; the response rate was 88 percent. Of the 71 nonrespondents, 22 had moved and tracing would have been required to locate them, 12 refused, and the remaining 37 could not be completed during the field period. No effort was made to convert those that refused the reinterview.

Interviewers were chosen from those who conducted the main study, but those selected were considered better than average by their supervisors. They were trained concerning the special requirements of the reinterview and were instructed to follow the same interviewing procedures and techniques followed in the original survey. The interviewer who conducted the original interview with a respondent was not permitted to conduct the reinterview with the same respondent.

The wording for the interview was kept exactly the same in the reinterview, although only a subset of the items was asked. A number of factors were considered in the selection of questions for the reinterview. The major factor was the requirement of examining the reliability of key questions for reporting and comparing over time. A second consideration was the utility of selecting a variety of questions in order to examine which types of questions are most subject to inconsistency. The third consideration was related to survey administration (e.g., some questions are connected with other questions and are difficult to replicate out of context). A copy of the reinterview survey appears in Appendix B.

The mode of reinterview was CATI -- the same as the original interview. The reinterviews were conducted in October and November, about 4 to 6 weeks after the original interview.

For key items, discrepancies between original responses and reinterview responses were subject to a reconciliation process. When responses differed, the respondent was first informed that a different response was recorded in the previous interview. Next, the respondent was asked which answer (the original or reinterview) was correct. Finally, the respondent was asked what he or she thought was the reason for the difference. (Summary tables on the reason for the difference are given in Appendix C.) All reconciliation was done after the completion of the entire reinterview so as not to influence results to subsequent questions. Furthermore, the interviewers did not know the original survey responses as they asked questions in the reinterview.

The use of CATI made it possible to conduct this type of study. Series of edit check screens were displayed to resolve differences between the original and the reinterview responses. While the answers were reconciled for purposes of the evaluation study, the data from the original interview were retained on the study analysis files.

Two models for measurement error are considered in this chapter. The relationships of the models to the reinterview study are explored. Later in the chapter, estimates of measurement error based on the models are developed and presented.

MEASUREMENT ERROR MODELS

Models of measurement errors have been proposed and refined by many researchers, with most following the general approach suggested by Hansen, Hurwitz, and Bershad (1961). Biemer and Stokes (1991) summarize many of these models and some of the theoretical development associated with them. Before discussing specific models for the RCG:91, an outline of the general idea may be useful.

Since it is difficult to directly estimate measurement error in a survey setting, models have been proposed to represent the most important structures of the error process. In essence, the models assume that the correct answer to a question may not actually be reported due to any number of sources of error. The measurement error model attempts to reflect the general nature of the errors, taking into account the data collection process. For example, a model might assume that in identical, independent replications of the data collection, the value reported would, on average, be the same as the correct value.

A measurement error model is useful only if it includes the major components of error. For example, if the model assumes that errors are independent, but they are actually highly correlated, then the estimates of the model parameters may be misleading. As a result, the model should be consistent with the design of study to ensure the validity of the model.

Simple Response Variance Model

The first model proposed for the RCG:91 assumes that the correct value differs from the observed value by an unobserved additive error term. The subscript t indicates that the response may be obtained on more than one occasion or trial (the original interview and the reinterview). The model is:

$$y_{ti} = \mu_i + \varepsilon_{ti} \tag{3.1}$$

where y_{ti} is the observed value at trial t for the ith respondent, μ_i is the unobserved correct value for the ith respondent, and ε_{ti} is the unobserved error at trial t for the ith respondent. To complete the specification of the model, we further assume:

$$\begin{aligned} E(\varepsilon_{ti}|i) &= 0 \\ Var(\varepsilon_{ti}|i) &= \sigma_i^2 \\ Cov(\varepsilon_{ti}\varepsilon'_{ti'}) &= 0 \ for \ i \neq i' \end{aligned} \tag{3.2}$$

We will refer to the measurement error model defined by (3.1) and (3.2) as model (3.2) for reasons that will be clear subsequently. Model (3.2) implies that there are no systematic biases in the estimates (the mean of the errors is zero) and the errors are not correlated. This latter means that the errors in one observation do not affect other observations and the errors across trials are uncorrelated and have identical first and second moments. There are ways in which this model can be modified to be more reflective of the design of the RCG:91. Some modifications will be examined later in this chapter and in the next chapter, after examining what can be done with this simple model.

Under the measurement error model, the ordinary measure of the precision of the estimate differs from the usual expression. The variance of a statistic, like a mean or proportion, can be written as:

$$Var(\bar{y}) = Var(\bar{\mu}) + Var(\bar{\varepsilon}) \tag{3.3}$$

The first term of (3.3) is the **sampling variance (SV)** of the estimate. The SV is the ordinary variance of the estimate if there is no measurement error. The second term of (3.3), often called the **simple response variance (SRV)** of the estimate, is the variability of the responses to the item averaged over conceptual repetitions of the survey under the same conditions.

Sometimes expression (3.3) gives the erroneous impression that the usual methods of estimating the variance of an estimate must be modified to account for the additional term. Hansen, Hurwitz, and Pritzker (1964) showed the ordinary estimate of the variance includes the measurement error. For example, in a simple random sample, the estimated variance of a mean can be decomposed as:

$$\frac{\sum^n (y_i-\bar{y})^2}{n(n-1)} = \frac{\sum^n \{(\mu_i-\bar{\mu})^2+(\varepsilon_i+\bar{\varepsilon})^2+2(\mu_i-\bar{\mu})(\varepsilon_i-\bar{\varepsilon})\}}{n(n-1)}. \quad (3.4)$$

Taking expectations, this expression reduces to (3.3).

Thus, if the assumptions of this measurement error model hold, the estimates from the survey will be unbiased and the estimated variance will include both the SV and the SRV. Despite this, it is still valuable to estimate the relative contribution of the SRV to the random error because the SRV can be reduced by different data collection methods (e.g., ways of phrasing the questions). If the SRV is a large fraction of the random error, then methods to reduce it can significantly reduce the errors in the estimates.

Estimators for the Simple Response Variance Model

The basic premise of model (3.2) is that the measurement errors are the same across sampled graduates and from one trial to the next. The original interview and the reinterview conform to this model in several ways. As discussed previously, the reinterview was conducted in much the same manner as the original interview. The mode was the same, the interviewers were taken from the same pool, CATI was used in both interviews, the same interviewer did not conduct the original and the reinterview for the same sampled graduate, and the same respondent was interviewed. These were design efforts to meet the assumptions of the model to the extent possible.

Clearly, the model fails to represent the actual conditions in some ways. One of the most probable causes of model failure is the correlation in responses between the original and the reinterview. The correlation may exist because the respondent recalls answers to the original question or is somehow influenced by the original survey. The reinterview was conducted 4 to 6 weeks after the original interview so that the respondent could not recall the original responses, but some conditioning is likely.[6]

Another reason this model might not be appropriate is the correlation between the responses of the sampled graduates that were conducted by the same interviewer. The interviewer contribution to the measurement error is the subject of the next chapter.

---

[6] It is worth noting that some characteristics might also change during the time between the original interview and the reinterview and this could result in differences in responses where both were correct. This situation was minimized because many items referred to a specific time period.

The other major reason for model failure involves the assumptions about the moments of the error term. For example, the error term may not have zero mean over replications of the survey and the error variances may be heterogeneous. These types of failures are likely to be of greatest concern for categorical data.

Despite its inherent limitations, model (3.2) can provide a useful approximation of the contribution of measurement error to the overall random error in the estimates from the RCG:91. To produce these estimates, the parameters of model (3.2) must be estimated from the original and the reinterview data. The trials are defined so that t=1 is the original interview and t=2 is the reinterview.

Gross Difference Rate

Under the assumptions of model (3.2), the response bias is defined to be zero and is not estimated. The SRV can be estimated by the **gross difference rate** (g), where g is:

$$g = \frac{1}{n}\sum (y_{1i}-y_{2i})^2. \qquad (3.5)$$

Biemer and Forsman (1992) show that the expectation of the g, under a more general model than (3.2) is given by:

$$Eg = SRV_1 + SRV_2 - \frac{2}{N}\sum \sigma_{t=1,2,i} + \frac{\sum (Ey_{1i}-Ey_{2i})^2}{N} \qquad (3.6)$$

Under model (3.2), this expectation reduces to:

$$E(g) = 2SRV_1, \qquad (3.7)$$

so the gross difference rate divided by 2 is an unbiased estimate of the SRV.

The proof of these results is based on simple random samples. To hold for more complex designs like the RCG:91, the estimators must be revised to include the sample weights. Appendix D provides details supporting the use of weighted measurement error estimators for complex sample designs.

In less technical terms, the gross difference rate is the weighted percentage of cases that were reported differently in the original and reinterview surveys. It is equal to the percentage of cases reported as having a characteristic in the original interview but not having it in the reinterview, plus the percentage of cases reported as not having the characteristic in the original interview but having it in the reinterview. That is, the gross difference rate is the ratio of the estimated number of cases misclassified in the original interview divided by the estimated total number of reinterview surveys.

Index of Inconsistency

A natural estimator of the proportion of the random error that is associated with measurement error is given by the **index of inconsistency**, (I):

$$I = \frac{SRV}{SRV+SV} \hat{=} \frac{g}{2s_1^2}, \tag{3.8}$$

where the denominator is estimated by the average of the ordinary variance estimates for the original and reinterview. Other estimators of the denominator of I are possible, but are not used in this evaluation.

With dichotomous variables, the estimators are often presented in a very simple table showing the original and reinterview estimates (or counts if the design is simple random sampling). Exhibit 3-1 shows the general format for reporting outcomes by the original interviews and reinterviews for dichotomous variables.

Exhibit 3-1. Interview by reinterview table

<table>
<tr><th rowspan="2"></th><th rowspan="2"></th><th colspan="2">Original interview</th><th rowspan="2">Total</th></tr>
<tr><th>Number of cases with characteristic</th><th>Number of cases without characteristic</th></tr>
<tr><td rowspan="2">Reinterview</td><td>Number of cases with characteristic</td><td>a</td><td>b</td><td>a + b</td></tr>
<tr><td>Number of cases without characteristic</td><td>c</td><td>d</td><td>c + d</td></tr>
<tr><td>Total</td><td></td><td>a + c</td><td>b + d</td><td>n = a + b + c + d</td></tr>
</table>

From tables formatted in this fashion, the gross difference rate and index of inconsistency take on very simple forms:

$$g = 100 \times \frac{b+c}{n} \tag{3.9}$$

$$I = 100 \times \frac{b+c}{2np(1-p)} \tag{3.10}$$

where p is $\frac{a+c}{n}$.

The estimators for g and I given above are just other ways of writing expressions (3.5) and (3.8) when the only two valid responses are zero and one.

For categorical variables with more than two response values, the expressions for the gross difference rate and index of inconsistency still can be written in forms that are simpler than expressions (3.5) and (3.8). For example, the gross difference rate is the sum of the off-diagonal elements of the original interview by reinterview table divided by the total for the table, expressed as a percentage. The index of inconsistency can be written as an average of the indices for the 2 x 2 sub-tables, often called the L-fold index of inconsistency. The U.S. Bureau of Census (1985) defines these terms more explicitly.

A different model can be formulated if the original response has a systematic error or bias that does not occur in the reinterview. We now explore the consequences of assuming that the second trial or the reinterview has less error than the original survey response.

RESPONSE BIAS MODEL

This new model retains the simple additive error structure of (3.1), but the assumptions on the error terms are different, since $\varepsilon_{2i} = 0$. The following results follow immediately from the assumptions about the error term:

$$\begin{aligned} E(\varepsilon_{ti}|i) &\neq 0 \quad for\ t = 1, \\ Var(\varepsilon_{ti}|i) &= \sigma_i^2 \quad for\ t = 1 \\ Var(\varepsilon_{ti}|i) &= 0 \quad for\ t = 2 \\ Cov(\varepsilon_{ti}, \varepsilon_{ti'}) &= 0\ for\ i \neq i' \end{aligned} \tag{3.11}$$

Note that in this model, the error term for first trial no longer averages to zero. The estimate based on the original interview could be subject to a **response bias**, where the bias is defined as:

$$\beta = \frac{1}{N}\sum(y_{1i}-\mu_i). \tag{3.12}$$

The subscript is omitted from the response bias because the response bias for the second trial is zero by assumption. This model will be called model (3.11).

In order to meet the conditions of model (3.11), the result from the reinterview should be free of measurement error. While this is not completely possible under the constraints of a reinterview, several different procedures have been proposed in the literature to obtain more accurate responses in the reinterview than were obtained in the original interview. These include using more experienced interviewers or supervisors, using improved data collection methods, using additional probing questions, and asking the respondent to reconcile the differences in responses.

For the RCG:91 reinterview, reconciliation of selected questions was chosen as the means of trying to obtain more accurate responses. Reconciliation was considered the most effective means of improving responses without adversely affecting the independent repetition of survey procedures needed to measure the SRV. Since it is very unlikely that the reconciled responses are actually error free, they can be used to identify the expected direction of bias, and the relative amount of bias, but cannot provide precise estimates of the size of the bias. Furthermore, the reconciliation process does not detect consistent errors made in both the original and the reinterview.

If the reconciled interviews are free of measurement error, the gross difference rate (computed as the difference between the original and the reconciled responses) no longer provides an unbiased estimate for the SRV. Using expression (3.6), it can be shown that the gross difference rate is an overestimate of the SRV (Biemer and Forsman, 1992). Therefore, the gross difference rate estimated using reconciled reinterview responses is an upper bound on the SRV. We will return to this point later in this chapter.

Net Difference Rate

Of course, the main reason for doing the reconciliation is to provide at least a rough guide to the size of the response bias. An unbiased estimate of the response bias under model (3.11) is given by the **net difference rate** (*ndr*), which can be written as:

$$ndr = \frac{1}{n}\sum (y_{1i} - y_{2i}). \tag{3.13}$$

This expression can be rewritten, using the terms from the interview-reinterview table as:

$$ndr = 100 \times \frac{c-b}{n}. \tag{3.14}$$

The net difference rate is the ratio of the net difference to the estimated total number of interviews. The net difference is the weighted difference between the total estimates for the variable of interest obtained from the original survey and the reinterview. The gross difference includes differences in any direction, and these differences may offset each other. The net difference is the non-offsetting part of the gross difference.

For items with multiple response categories, the net difference is defined as the number of cases above the main diagonal minus the number of cells below the main diagonal. Items which are measured in constant, linear units (e.g., number of hours) and are symmetric about the diagonal can be treated in much the same manner as items with only two categories.

While expression (3.13) is valid for both quantitative and dichotomous variables, it is less justified when the responses are categorical, unordered data. For example, for a variable such as race, which takes on the value 1, 2, or 3 corresponding to white, black, and Native American, this expression for the net bias actually weights the responses. In this case, the difference between response categories 1 and 3 would result in a larger contribution to the net bias than the difference between 1 and 2. Since these are unordered responses, this approach is questionnable. Because of this, the net difference rate for the few categorical, unordered response variables were computed differently. The net difference rates were computed without weighting the responses. In other words, the difference between white and black would count the same as the difference between white and Native American. However, for these types of unordered measures the net difference rate is more of a general indicator of offsetting error than a direct measure.

While the net difference rate computed based on the reconciled responses can be used to estimate the expected direction and magnitude of response bias, it does not have the same properties when computed using the unreconciled responses. In fact, it can be shown that the unreconciled net difference rateis an indicator of how well the reinterview meets the uncorrelated assumption of model (3.2). A high net difference rate suggests that the reinterview may not have replicated the original survey very well. This could result in the gross difference rate being an overestimate of the SRV.

Special Case for Dichotomous Variables

In discussing model (3.2), we mentioned some of the problems of using this model with categorical variables. We expand on that discussion below with particular attention to dichotomous variables that take on the value of one if the sampled unit has the characteristic and zero otherwise.

With a dichotomous variable, the conditions on the moments of the model (3.2) can be written in terms of the probabilities of misclassifying the sampled graduate (falsely classifying the graduate as having or not having the characteristic). Biemer and Forsman (1992) show that both the response bias and the SRV are functions of these probabilities of misclassification and the proportion of the graduates that have the characteristic. They show that the response bias is zero only under special conditions. For example, the response bias is not equal to zero when the misclassification errors (false positives and false negatives) are equal, except for characteristics held by exactly 50 percent of the population.

These results have implications for the interpretation of the response bias and the SRV for dichotomous variables. The assumption of zero response bias in (3.2) does not mean that the probability of misclassification is the same in both directions. Rather, it means the number of sampled graduates erroneously classified as having the characteristic will, on replications of the survey, equal the number of graduates erroneously classified as not having the characteristic.

The SRV is still estimated unbiasedly by half of the gross difference rate, but it does not directly measure the probabilities of misclassification. Thus, the index of inconsistency, I, is an estimator of the impact of misclassification errors on the estimates rather than a direct measure of the misclassification probabilities. The Appendix in the U.S. Bureau of the Census report (1985) describes these issues in more detail and gives some tables to demonstrate these points.

While these issues affect the interpretation of the measurement error estimators, the model is not invalidated by them. For example, it is easy to show that the zero expectation assumption in (3.2) can be relaxed without impacting the estimators. A new model that includes a nonzero bias can be transformed simply into the form given by (3.2) provided the bias is the same from trial to trial. Under this model with a constant bias across trials, the response bias cannot be estimated from the unreconciled reinterview and original interview responses, since the bias term is contained in both observations. However, the constant bias does not have any effect on the unbiasedness of the gross difference rate as an estimator of the SRV. Of course, the usual estimate of the variance of the estimate does not capture this constant bias, and it, therefore, underestimates the mean square error of the estimate.

These results indicate that the proposed estimators of measurement error are valid under the models presented for both quantitative and categorical variables. Special care must be exercised in the interpretation of these estimators for categorical data. These measures are applied below using the original interview, the reinterview, and the reconciled reinterview data from the RCG:91.

FINDINGS

Table 3-1 presents the gross and net difference rates and indices of inconsistency for key items. Both reconciled and unreconciled results are presented. Table 3-2 presents unreconciled results for items included in the reinterview that were not reconciled. The items in Table 3-2 were selected for inclusion in the reinterview as examples of questions that might have the potential for measurement errors for various reasons such as sensitivity, date recall problems, and question complexity. The sample size varies from item to item because of skip patterns in the interviews.

As noted in the sections above, the primary focus of the RCG:91 reinterview study was to measure the random component of measurement error using the gross difference rate and the index of inconsistency, based on the unreconciled data. The net difference rate based on the reconciled data is used as a gross measure of the direction and magnitude of the potential response bias, but this measure is limited. Other measures, such as the net difference rate based on unreconciled data, are presented for completeness and as checks of the validity of the model. These uses are summarized in Exhibit 3-2.

Some rough rules of thumb for interpretation have been suggested for using the index of inconsistency as an estimator of the impact of measurement error on the estimates (U.S. Bureau of Census, 1985). These rules are most applicable when the estimated characteristic is between 20 and 80 percent. The rules are, if the index of inconsistency is:

- Less than 20, the impact of measurement error is low;
- Between 20 and 45, the impact of measurement error is moderate; or
- Greater than 45, the impact of measurement error is high.

Table 3-1. Gross and net difference rates and index of inconsistency for reconciled and unreconciled key items from the RCG:91 reinterview

Key item		Reinterview sample size	Population estimate (percent yes or mean)	Reconciled			Unreconciled				
				Gross difference rate[1]	Net difference		Gross difference rate[1]		Net difference rate[2]	Index of inconsistency[3]	
					Rate	Selected standard errors	Rate	Selected standard errors		Index	Selected standard errors
Employment											
Q23R	Was R working for pay in reference week	510	84	2.31	.73	.41	4.35	.71	.50	17.19	3.41
Q24R	If not working was R looking for work in reference week	66	34	15.12	.14	4.08	26.41	4.49	4.57	58.78	11.03
Q25R	If not working was R available for work in reference week	67	40	5.09	1.23	2.05	5.76	2.18	.56	11.88	4.52
RSOC	If working, what was R's occupation	423	NA	*	*	*	2.56	.65	.18	2.83	.71
Q32R	If working, was job full time or part time	423	13	5.14	1.24	.71	6.79	.97	.39	28.60	3.84
Q34R	What kind of employee was respondent	423	NA	3.32	.10	.52	4.99	.74	-1.57	10.33	1.47
Q40R	If working, was R working at second job	423	13	3.45	.21	.71	6.15	.95	1.72	29.55	4.19
Additional Education											
Q11R	Did R apply to additional schools . . .	512	33	6.49	-1.97	.80	9.15	1.03	.21	20.25	2.34
Q12R	Has R enrolled since receiving degree	512	34	6.91	-2.73	.97	8.87	1.01	-1.56	19.34	2.24
Q15R	Is R still enrolled . . .	163	62	2.52	1.87	1.06	3.92	1.34	3.26	8.73	3.01
Q23AR	Did R have assistantship or work study	163	13	4.52	1.73	1.25	6.25	1.33	1.93	24.57	5.82
Teacher status											
Q50R	Is R eligible to teach at any level	512	17	1.43	0.13	.39	2.64	.52	.71	10.12	1.97
Q53R	Is R certified to teach at any level	512	16	1.26	-0.14	.33	1.26	.33	-.14	4.90	1.29
XNQTR	Newly qualified to teach status	512	13	*	*	*	4.32	.49	-1.72	21.08	2.79
Q62R	Before degree was R employed as teacher .	512	10	*	*	*	3.96	.69	1.28	63.0	7.34
Q68R	Principal job as school teacher	132	12	12.57	-11.57	2.43	14.55	2.77	-9.59	30.43	5.56
Income											
Q39	Annual rounded to $2,000	398	$24,000	14.97	3.13	1.61	19.77	1.63	1.16	23.23	1.93
Q39	Annual rounded to $4,000	398	$24,000	8.54	2.75	1.29	10.53	1.14	1.43	12.65	1.37
Q39	Annual rounded to $5,000	398	$24,000	6.51	1.02	1.03	8.20	.95	-0.43	10.18	1.16
MAJ87	Major field of study (12 categories)	512	NA	*	*	*	3.17	.53	.60	NA	NA
Q67Date	Date R started to teach	109	NA	15.17	6.02	2.95	27.52	3.16	10.21	32.42	3.79
Q52R	Date became eligible to teach	113	NA	6.55	-3.96	1.71	17.91	2.69	.40	46.48	6.80

NA - Not applicable or not calculated.

* Item not reconciled.

[1]The gross difference rate is the weighted percentage of cases that were reported differently in the original and reinterview surveys.

[2]The net difference rate is the ratio of the net difference to the estimated total number of interviews. The net difference is the weighted difference between the total estimates for the variable of interest obtained from the original survey and the reinterview.

[3]The index of inconsistency is the ratio of the variance of the response errors to the total variance of the measure.

SOURCE: U.S. Department of Education, National Center for Education Statistics, *1991 Recent College Graduates Survey.*

Table 3-2. Gross and net difference rates and index of inconsistency for selected items from the RCG:91 reinterview classified by type of questions

Selected item		Reinterview sample size	Population estimate (percent yes or mean)	Unreconciled**				
				Gross difference rate[1]		Net difference rate[2]	Index of inconsistency[3]	
				Rate	Selected standard errors		Rate	Selected standard errors
Sensitive/social desirability								
Q10	Grade point average	506	NA	22.62	1.42	-2.87	33.17	2.02
Race/ethnicity								
Q95R	Race	500	NA	1.05	.44	.17	7.45	3.11
Q94R	Hispanic origin	510	4%	.46	.27	-.26	9.78	5.81
Annual teaching income (see also Table 3-1)								
Q87CR2R	Rounded to $2,000	59	$19,400	9.33	2.48	3.46	10.63	2.84
Q87CR4R	Rounded to $4,000	59	$19,400	9.33	2.48	3.46	12.87	3.38
Q87CR5R	Rounded to $5,000	59	19,400	7.70	2.45	5.09	11.63	3.72
Date Questions								
Q52R	When did R first become eligible (categories)	113	NA	17.91	2.69	.40	46.48	6.80
Q55YYR	Year R was certified to teach	119	1,990	16.56	2.95	-9.16	35.64	6.37
Q55Date_R	Month and year R was certified to teach within 2 months	115	NA	20.50	2.75	.45	22.91	3.21
Q55DATE	Date R became certified, exact month and year	115	NA	53.61	3.82	2.77	58.63	3.98
Q87AR	Number of months teaching contract	63	10	35.06	4.32	1.82	50.84	6.64
Q87BR	Number months paid for teaching	63	12	17.55	3.02	1.80	30.39	4.47
Q38R	Hours per week working	412	31-40	31.60	2.35	-5.77	--	--
Anticipated ambiguous/complex questions								
Q106R	Sources of support for financing degree	4,070*	NA	4.67	--	.75	16.83	--
Q110A1.E1	Types of federal financial aid received (overall)	1,258*	NA	12.26	--	2.42	26.42	--
Q110A1.E1	Did R get specific types of federal aid in 89-90	347	NA	16.85	--	-3.79	35.25	--
Q58	Fields eligible to teach	2,642*	NA	11.42	--	-1.74	33.68	--
Q58Elem	Fields eligible to teach elementary	2,022*	NA	13.63	--	-3.51	36.41	--
Q58Nonelem	Fields eligible to teach nonelementary	620	NA	5.23	--	3.05	23.88	--
Q59	Fields in which certified to teach all	2,733*	NA	8.55	--	-1.39	29.50	--
Q59Elem	Fields in which certified to teach elementary	2,136*	NA	10.21	--	-1.95	32.09	--
Q59Nonelem	Fields in which certified to teach nonelementary	597*	NA	2.81	--	.55	15.97	--
Q71	Fields in which teaching all	2,838*	NA	6.30	--	-1.48	46.79	--
Q71Elem	Fields in which teaching elementary	1,056*	NA	11.42	--	-4.72	92.50	--
Q71Nonelem	Fields in which teaching nonelementary	1,782*	NA	3.25	--	.43	23.03	--
Opinion/perception questions								
Q13R	Best reason for not applying for additional school	269	NA	24.14	2.02	-7.76	37.82	3.52
Q42R	Was college degree required for job	419	56	11.68	1.04	-2.30	23.74	2.04
Xrelate	Was job related to degree	512	NA	6.91	1.04	2.51	15.74	2.41
Xpotent	Was there career potential	512	NA	10.28	1.04	.61	23.17	2.43
Q65R	Main reason did not apply for teaching	362	NA	4.03	.88	.43	91.04	18.05
Teaching status								
Q85R	Was teaching assignment full time	87	88	10.12	1.17	1.89	34.61	3.91
Q87DR	Any summer employment besides teaching	63	34	8.98	2.36	6.30	18.62	4.87
Q87R	Have teaching contract or some other arrangement	84	81	1.81	.90	-1.81	4.58	2.26
Change from time of interview possible								
Q12R	Enrolled since degree	512	34	8.87	1.01	-1.56	19.34	2.24
Q66R	Has respondent taught any grade since degree	161	16	18.84	2.87	-17.19	39.18	5.55

*Subquestions for multipart question combined.

**Most of these questions were not reconciled in the reinterview. (Grade point average was reconciled: the gross difference was 13.63; the net difference was -1.08; the index of inconsistency was 19.98. Best reason for not applying for school was reconciled: the gross difference rate was 14.73; the net difference rate was -5.43, and the index of inconsistency was 23.08.)

--Not calculated.

NA = Categorical or combined data.

[1]The gross difference rate is the weighted percentage of cases that were reported differently in the original and reinterview surveys.

[2]The net difference rate is the ratio of the net difference to the estimated total number of interviews. The net difference is the weighted difference between the total estimates for the variable of interest obtained from the original survey and the reinterview.

[3]The index of inconsistency is the ratio of the variance of the response errors to the total variance of the measure.

SOURCE: U.S. Department of Education, National Center for Education Statistics, *1991 Recent College Graduates Survey.*

Exhibit 3-2. Uses of reinterview statistics, by type of reinterview responses

Statistic	Type of reinterview responses	
	Unreconciled	Reconciled
Gross difference rate	Measure of random error (simple response variance)	Model diagnostic
Net difference rate	Model diagnostic	Measure of systematic error (response bias)
Index of inconsistency	Ratio of simple response variance to total random error	

General Comments on Errors

The gross difference rates and indices of inconsistency of the key items in Table 3-1 generally indicate low to moderate random measurement error. The only key items in Table 3-1 that could be considered to be subject to high random measurement error are the one that asks unemployed persons if they were looking for work (Q24R) and the one that asks if the graduate was employed as a teacher before getting the degree (Q62R).

The gross difference rates and indices of inconsistency are somewhat larger for the items in Table 3-2, many of which were selected because of anticipated high error (such as sensitive, date, and ambiguous/complex items). Even for these items, the measurement errors seem moderate, with 32 percent having low indices, 50 percent having moderate indices, and 17 percent having high indices, using the rule of thumb given above.

Figure 3-1 shows the mean gross difference rates for groups of variables from Tables 3-1 and 3-2. The figure shows the gross difference rates are generally small, with the exception of the rates for items that asked about dates.

If we assume the reconciled value is the correct response and apply model (3.11), we can estimate the response bias for the key items in Table 3-1. The net difference rate is small for almost all of the key items, with only two items with a net difference rate of greater than 5 percent. Only one of these two items, the question about the respondent's principal job as a teacher (Q68R), was significantly different from zero when the sampling error of the net difference rate was taken into account.

These findings indicate that either the response bias is small for the items that were reconciled or the reconciled reinterview did not result in significantly reducing any of the bias that may have been associated with the items. It is impossible to disentangle this possible model failure from the lack of a response bias. As a result, a prudent statement is that the reinterview did not provide evidence of significant response bias for nearly all of the key items in the RCG:91.

Figure 3-1. Mean unreconciled gross difference rates for selected groups of items

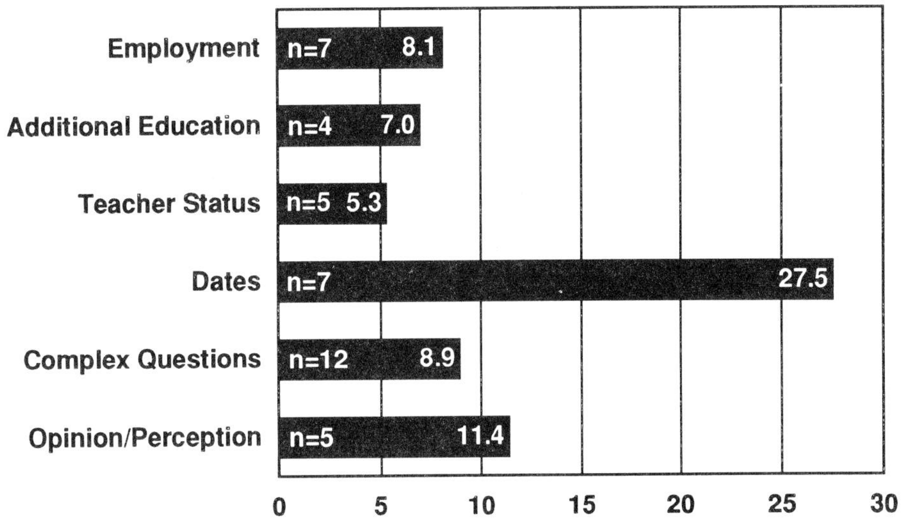

NOTE: Represents mean of data in selected categories taken from Tables 3-1 and 3-2.

Errors by Types of Variables

In addition to the items with larger than average error estimates, some general observations about types of variables are possible. Among the RCG:91, questions about the status of the graduates, including items such as whether the respondent was working, type of employee, eligibility to teach, certification to teach, newly qualified to teach status, occupation, and major field of study, have low measurement errors with unreconciled gross difference rates under 5 percent, reconciled net difference rates of under 2 percent, and indices of inconsistency of under 20 percent.

Annual income, a key variable on the RCG study, had low net difference rates and moderate unreconciled gross difference rates. The index of inconsistency for income was 23 percent, when income was rounded to $2,000.

The items on additional education, which had some potential to change with time, had gross difference rates of under 10 percent, reconciled net difference rates of 2 to 3 percent, and indices of inconsistency between 9 and 20 percent.

The opinion questions on the career potential and the necessity of college degree for the job had unreconciled gross difference rates of about 10 to 12 percent, net difference rates of about 1 to 2 percent, and indices of inconsistency of just over 20 percent. The question on whether the job was related to degree had a gross difference rate of 7 percent, a net difference rate of 3 percent, and inconsistency index of 16 percent.

The questions on the sources of financial support for attending college were at the end of the survey and asked for some detailed information on types of federal aid received in a specific year. These items were anticipated to be more ambiguous or complex than many of the other items in the survey.

These questions had moderately low measurement errors. The overall questions on sources of support had a gross difference rate of 5 percent, a net difference rate under 1 percent, and an index of inconsistency of 17 percent. The questions for the subgroup receiving federal aid on the types of aid received at any time for the degree and in the specific year of 1989-90 also had moderate measurement errors.

While questions on whether the respondent was certified, eligible, or was a newly qualified teacher in Table 3-1 had very low measurement error, as indicated by the reinterview results, questions on the specific fields in which the respondent was eligible, certified, or teaching in Table 3-2 had moderate measurement errors. Differences were higher for eligibility and certification questions than for teaching fields. Overall, about 10 to 12 percent of the possible fields had different responses on the reinterview than the original for the eligibility and certification questions; for the teaching fields, the estimate was 6 percent.

Difference rates were considerably higher for elementary than nonelementary teachers. For example, the unreconciled gross difference rate for certification in specific fields was 3 percent for nonelementary and 10 percent for elementary. These results point to the fact that the questions, while working well for secondary teachers, could be improved for elementary teachers.

Items with Large Measurement Errors

As mentioned earlier, the question about whether a person not employed in the reference week was looking for work had an unreconciled gross difference rate of 26 percent and an index of inconsistency of 59 percent. This question was only asked of a small subgroup of the population, those currently not working. The question has some potential for recall problems since it was linked to a specific week and required the respondent to have taken certain specific actions, such as answering an ad or making a specific job-seeking phone call. These types of questions are more suspectable to measurement error than other questions that make direct inquires about the current status of the graduate.

The survey also included a number of date questions in which the graduate was asked to provide the exact month and year. These questions had the highest gross difference rates. When asked the exact month and year they became certified to teach, more than half the respondents gave a different response on the reinterview (gross difference rate of 53 percent). However, about 80 percent of respondents were within 2 months of the original response on the reinterview, and 83 percent reported the same year as on the original interview (21 and 17 percent gross difference rates, respectively). In effect, the problem with exact dates is overstated because the dates are treated as categorical, when they are really more continuous in nature. For the one question (Q52R) in which a range of dates was given, the gross difference rate was only 18 percent. The index of inconsistency for this question was high (46 percent), due in part to the fact that most of the responses clustered into one of the categories.

A question that proved especially ambiguous for respondents concerned the number of months of their teaching contract. About one-third of respondents gave a different number of months on the reinterview than the original survey. For some, the difference was between 9 and 10 months; for others, the difference was between 12 months and 9 or 10 months. The number of months paid for teaching was less problematic.

The question on grade point average could be considered somewhat sensitive and also has potential recall problems. Overall about 23 percent of the reinterview respondents had a difference in categorization from the original interview, and the index of inconsistency was moderate (33).

Comparison with Other Reinterview Studies

The reinterview findings from RCG:91 are consistent with the conclusion of other reinterview studies that items asking for factual and status information have lower response variability than questions asking for opinions, or more complex responses (National Center for Education Statistics 1984; Bushery, Royce, and Kaspryzk, 1992). For example, in RCG:91 as in other surveys, very low levels of variability were found for status questions such as race/ethnicity and teacher certification, and relatively higher levels were found for the few opinion questions such as reasons for not applying for additional education.

Other studies have also found that items asking for recent or current information have lower variability than those that are retrospective or ask for future expectations. It has also been found that the more open-ended the response choices, the more specific a date requested, and the higher the number of response categories, the greater the variability. Some have recommended on the basis of reinterviews that certain questionnaire categories be combined and that more direct yes/no questions be developed (Bushery, Royce, and Kaspryzk, 1992).

The moderate level of variability for the income items on RCG:91 was fairly consistent with that found in other studies. For example, the SASS survey found gross difference rates of 9.6 percent for public school teachers and 13.9 for private school teachers for annual salary when grouped into four broad categories and rates of about 20 percent when grouped in smaller (5 percent) categories of income. The RCG:91 rate of 9.3 percent when teaching income was rounded to plus or minus $2,000 is consistent with this result.

Within the RCG:91, less variability was found for teacher income when asked in the form of annual income than was found for the general salary question in which respondents were able to choose the unit in which to respond (hourly, weekly, monthly, yearly). The gross difference rate for this question was 16.0 percent when income was rounded to plus or minus $2,000.

Checks on the Model

In the previous section, it was noted that high net difference rates based on unreconciled data could be used as a diagnostic of one of the assumptions of model (3.2). If the unreconciled net difference rates are large, the result could be the gross difference rate overestimating the SRV.

Of the 57 items appearing in Tables 3-1 and 3-2, there are only 9 items that have net difference rates significantly greater than zero at the 95 percent confidence level, when computed based on the unreconciled reinterview data. This is more than the expected number (3), but it does not indicate a gross failure of the assumption of model (3.2).

Another model check proposed by Biemer and Forsman (1992) is to compare the unreconciled gross difference rate with two times the reconciled gross difference rate. If the assumptions of model (3.2) hold, then two times the reconciled gross difference rate should be greater than or equal to the unreconciled gross difference rate.

For all of the estimates in Table 3-1, except one, this upper bound is satisfied. This is another indicator that the assumptions of model (3.2), while not representing all of the aspects of the survey conditions, are generally consistent with the data.

A byproduct of the reconciliation process can also be used as a check on the model assumptions. When there was a discrepancy between the response on the original and reinterview, the graduate was informed of the discrepancy, asked to identify the correct answer, and then asked for a reason for the discrepancy. The reasons for discrepancies provide insights for revising the items for future surveys and are addressed separately in Appendix C.

If the reinterview is uncorrelated with the original interview, as assumed under model (3.2), then the number of original and reinterview errors should be roughly equal. Table 3-3 shows the distribution of responses to the resolution of discrepancies. Overall, graduates said that the original answer was correct for 43 percent of the discrepancies, and that the new (reinterview) answer was correct for 47 percent of the discrepancies.

This distribution changes when the income items (Q37 and Q39) are excluded. For income, small differences due to rounding and differences in the reporting unit (year, month, week, day, or hour) were included as discrepancies. If the graduate indicated that the responses were actually the same, the interviewer was instructed to choose the category "original answer correct." Therefore, the "original answer correct" category is slightly inflated for the income items, and a more accurate distribution may be obtained when the income items are excluded. When income items are excluded, graduates said that the original answer was correct for 39 percent of the discrepancies, and that the reinterview answer was correct for 50 percent of the discrepancies.

The differences between the percentage of discrepancies in the two categories (original answer correct and new answer correct) are small, tentatively supporting the finding that the reinterview was relatively successful in producing an uncorrelated replication of the original survey. Of course, this measure only relates to those responses that were different. The proportion of response with different answers is also affected by any correlation, but this effect is confounded with the level of the measurement error.

Table 3-3. Resolution of response discrepancies

Resolution of discrepancies	Total		Excluding income items	
	Number	Percent	Number	Percent
Total discrepancies . . .	899	100%	671	100%
Original answer correct	390	43	263	39
New answer correct . . .	424	47	335	50
Neither answer correct[1] .	25	3	17	3
Situation has changed[2] .	40	4	40	6
Don't know	20	2	16	2

[1]This resolution was only applicable for questionnaire items with more than two response categories. This includes Q10, Q13, Q28, Q34, Q36, Q37, Q38, Q39, Q52, and Q67. Among the cases where "neither answer correct" was applicable, it was chosen 25 of 656 times (4 percent).

[2]This resolution was only applicable for questionnaire items where it was possible for the situation to change. This includes Q11, Q12, Q13, Q15, Q50, and Q53. Among the cases where "situation changed" was applicable, it was chosen 40 of 200 times (20 percent).

NOTE: Details may not add to totals due to rounding.

SOURCE: U.S. Department of Education, National Center for Education Statistics, *1991 Recent College Graduates Survey*.

IMPLICATIONS AND RECOMMENDATIONS ON MEASUREMENT ERRORS

In this chapter, two models were presented to represent measurement errors, estimators of the parameters of those models were provided, and the findings were applied using the data from the reinterview study. Some checks on the assumptions of the models were also examined, and they indicated that the models were moderately consistent with the data.

The overall conclusions of the study show that even though measurement errors were an important source of error in the RCG:91, the estimates from the survey were not greatly distorted by these errors. The gross difference rates, as measures of the random component of measurement error, were relatively small for most of the variables in the reinterview study, indicating that the reports from the sampled graduates were consistent.

The indices of inconsistency provided estimates of the impact of the measurement error on estimates. These estimates were generally moderate, implying that improvements in questionnaire wording and construction might help to reduce measurement errors. Specific questions were also identified that require more substantial revisions. When using these specific questions, data users should be cautious in their statements.

The estimates of response bias were more tenuous. These estimates, the net difference rates computed from reconciled reinterview data, depend on model assumptions that are not as consistent with the reinterview procedures and cannot be examined from the data. Despite these problems, the fact that nearly all of the estimates had small or insignificant net difference rates should be encouraging to users.

The random component of the measurement error (the SRV) was shown to be included in the usual estimate of the variance of the estimate. This result, combined with the general overall finding that the checks support the assumptions of model (3.2), imply that the usual methods of variance estimation account for both sampling and measurement error. Thus, confidence intervals and significance tests computed using the appropriate estimates of the sampling errors of the estimates should produce valid statements.

Of course, the models presented in this chapter do not account for all sources of measurement error. In the next chapter, the interviewer as a source of measurement error is considered. An important result is that systematic errors, such as those that can result from interviewers or coders, are not generally incorporated in the estimates of the variance of the estimates. Before drawing final conclusions these effects should be investigated.

4. NONSAMPLING ERROR FROM MEASUREMENT ERROR: INTERVIEWER MEASURES

In the previous chapter, models for measurement error were presented that assumed there was no correlation between the responses of sampled graduates. This assumption may not be valid for the RCG:91 because the telephone interviewers generally conducted many interviews with sampled graduates. If the interviewers brought methods of asking questions and recording responses that were idiosyncratic, these systematic differences may have resulted in measurement errors that were not recognized by the models employed thus far.

The same result could also hold for any other source of errors that caused the observations across sampled graduates to be correlated. For example, coders of verbatim responses to specific items could play the same role as interviewers. However, in this evaluation, attention is restricted to interviewers because they were an important potential source of measurement error in the collection of data for all of the questions in the study.

The most theoretically sound method for estimating the contribution of interviewers to measurement error is to design the data collection procedures with this goal. One design for doing this is called an interpenetrating sample design (e.g., see Mahalanobis, 1946). Basically, this method assigns the sampled units to the interviewers randomly. The operational features of an interpenetrating design for a central telephone survey are very difficult and expensive to implement without negatively impacting on response rates (Groves and Magilavy, 1986). For example, many interviewers may only be available to work during certain hours of the day. If a sampled graduate cannot be reached during the times the interviewer works, then the graduate could end up as a nonrespondent under a strict random assignment design. Since this negative impact on the response rates was not acceptable in the RCG:91, an interpenetrating design was not used.

An alternative approach is used instead for estimating the interviewer contribution by developing a model that explicitly recognizes the nonrandom assignment of the sample to the interviewers. We examine this model and its consequences in some detail in this chapter, following the general organization of the previous chapters. We begin by presenting the procedures for the interviewer effects analysis, present a model for measurement errors from these data, discuss methods for estimating the parameters of the model, and apply those methods to the RCG:91 data. The final section of this chapter covers the implications of these findings for data users and future surveys.

PROCEDURES FOR THE INTERVIEWER EFFECTS ANALYSIS

In many studies, the interviewers' input in asking questions, probing for responses, and recording those responses has been found to have a large impact on the error of the estimates. For examples, see Hanson and Marks (1958), Kish (1962), Bailar (1968), and Pannekoek (1988). In the RCG:91, the average number of sampled bachelor's recipient interviews conducted by an interviewer was in excess of 100. Since the impact of the interviewer contribution to measurement error increases with the number of interviews conducted (discussed later), the large interviewer case load for the RCG:91 makes this source of error potentially very important.

Analytic methods were used to account for the nonrandom assignments of the sample to the interviewers. Below, the procedures used to prepare the RCG:91 data set for the analysis are presented, along with a brief discussion of the reasons for preparing it in this fashion. The rationale for these procedures will be clearer when the model is examined later in the chapter.

Data Used in the Analysis

The data used for the analysis were restricted to bachelor's degree recipients. Interviewer effects also affect the estimates for master's degree graduates, but the impact for these graduates should be considerably smaller because the average interviewer load was less than 15 for master's degree graduates.

The full RCG:91 data set contained completed interviews for 12,888 bachelor's degree graduates. For this analysis, interviews were deleted from the full data set if they were assigned to specific interviewers or groups of interviewers with special training or skills, such as in refusal conversion or language problems. Furthermore, cases that were missing key items were dropped. This reduced data set contained 12,236 completed interviews.

Another problem occurred because the procedures available in standard statistical packages for computing estimates of the components of variances for the types of models proposed below do not adequately account for differential sampling weights. Even though accounting for weights in this type of analysis is often not critical (e.g., the unweighted reinterview estimates shown in the previous chapter were nearly identical to the weighted estimates), a scheme was used to eliminate this and other related potential problems.

A sample of the graduates was selected from the 12,236 observations so that the analysis could be conducted from an approximately self-weighting file. To implement this, all of the cases with weights greater than or equal to the 90 percentile of the weight distribution were included in the sample. For the remaining cases, the probability of selection for each case was set equal to the weight of the case divided by the weight at the 90 percentile of the weight distribution. This probability was compared to a randomly drawn number between zero and one for each case. If the random number was greater than the probability, the case was dropped from the analysis file. The result of this sampling was a self-weighting analysis file with 8,761 cases.

The only other manipulation of the file for this analysis involved dropping cases from the individual runs if the response for the particular question was imputed (all missing values were imputed in the RCG:91). As noted in Chapter 2, there was very little missing data so this restriction had little impact on the sample size.

INCORPORATING INTERVIEWERS IN THE MEASUREMENT ERROR MODEL

The model of measurement errors presented in the previous chapter has been extended to include the interviewer as a source of error, following Hansen et al. (1951). This approach is also summarized in the review article by Biemer and Stokes (1991). In simplified terms, the study of interviewer effects assumes that interviewers are a random sample from an infinite pool of possible interviewers. The goal of the analysis is to determine if the interviewers bring specific biases or effects to the interviewing task. If they do have systematic effects, then the purpose is to relate these effects to the estimation of the reliability of the survey estimates.

If the interviewers are an important source of measurement error, then the model represented by equations (3.1) and (3.2) is not appropriate. A modified model that explicitly includes the potential contribution of interviewers to the measurement error is given by:

$$y_{ji} = \mu_i + \beta_j + \varepsilon_{ji} \qquad (4.1)$$

where βj is the systematic error associated with interviewer *j*. The interest in this model lies in inferences to the population of interviewers, not the specific interviewers in the study. Thus, the interviewer effect is a **random effect.** It is also worth noting that the subscript for the repeated trials has been dropped in the model, but the conceptual framework for the error distribution is based on the outcomes of repeated surveys conducted under the same conditions.

The assumptions for this model are given by:

$$\begin{aligned} E(\varepsilon_{ji}|i) &= 0 & & \\ Cov(\varepsilon_{ji,}\varepsilon_{j'i'}) &= 0 & & \text{if } j \neq j' \\ &= \rho^*\sigma^2 & & \text{if } j = j',\ i \neq i' \\ &= \sigma^2 & & \text{if } j = j',\ i = i' \end{aligned} \qquad (4.2)$$

This model allows for a correlation between the observations conducted by the same interviewer, but assumes there is no correlation between interviewers and no correlation between the actual value and the error term.

The variance of a mean or a proportion becomes more complex due to the correlation between interviews conducted by the same interviewer. It can be written as:

$$V(\bar{y}) = V(\bar{\mu}) + V(\bar{\beta}) + V(\bar{\varepsilon}) + Cov(\bar{\mu},\bar{\beta}) + Cov(\bar{\mu},\bar{\varepsilon}) + Cov(\bar{\beta},\bar{\varepsilon}). \qquad (4.3)$$

If we assume the interviewer error is uncorrelated with the true value for the unit along with the assumptions of model (4.2), this expression can be reduced to:

$$V(\bar{y}) = V(\bar{\mu}) + V(\bar{\varepsilon}) + V(\bar{\beta}) + Cov(\bar{\beta},\bar{\varepsilon}). \tag{4.4}$$

The first two terms of this expression are the sampling variance (SV) and the simple response variance (SRV), as defined in the previous chapter. The last two terms are collectively called the **correlated component of response variance (CC).**

The rationale for this name follows by re-expressing the CC as:

$$CC = \frac{\sum m_j(m_j-1)\rho^*\sigma^2}{n}, \tag{4.5}$$

where m_j is the number of interviews conducted by interviewer j.

The **intra-interviewer correlation coefficient**, a commonly reported measure of the impact of the interviewer on the variance of the estimate, is given by:

$$\rho = \frac{\rho^*\sigma^2}{SRV+SV} = \rho^* I, \tag{4.6}$$

where I is the index of inconsistency defined in the previous chapter.

If the interviewer work loads are all equal to m, then expression (4.4) can be written as:

$$V(\bar{y}) = (SV+SRV)(1+(m-1)\rho). \tag{4.7}$$

Since the intra-interviewer correlation coefficient is nonnegative, the impact of any systematic error due to interviewers is to increase the variance of the mean. Note that even if the correlation is small, the impact on the variance of the mean can be large if the interviewer sample size is large. For the RCG:91, the variance of the estimate would be three times as large if the correlation was only 0.02, since the average interviewer work load was over 100.

An important difference between this error model and that represented by (3.2) is in the estimation of the precision of the estimates. If model (4.2) holds, then the ordinary estimate of the sampling variance does not include the contribution of the errors due to interviewers. The ordinary estimate of the variance does not incorporate the interviewer effect and, therefore, underestimates the variability of the estimates.

The underestimation of the variance of the estimates can pose a serious problem for data users, since it results in confidence intervals and significance tests that are nonconservative. The extent of this problem is considered after methods for estimating the intra-interviewer correlation are discussed below.

Estimators of Intra-interviewer Correlation

Kish (1962) proposed using the usual ANOVA table to estimate the intra-interviewer correlation component for an estimated mean from a survey. One of the problems with that approach for the RCG:91 is the lack of randomization of the cases to the interviewers, which is a basic assumption for the application of ANOVA methods. To better understand the rationale for the proposed solution to this problem, a brief description of the method of assignment of cases in the RCG:91 is needed.

The interviews were assigned to interviewers using the Westat system of scheduling cases in a centralized telephone facility. Under this scheduling system, the vast majority of cases are assigned systematically to the next available interviewer according to a priority scheme that is independent of the interviewer. In other words, the scheduling may depend upon the calling history of the case (in terms of days and times it has been previously called), but the characteristics of the interviewer are not used in the assignment procedure.

There are important exceptions to this general rule. Groups of interviewers may be assigned to special categories of cases, such as refusal conversions and language problem cases. If a case is placed in one of these categories, then only interviewers who are specially trained for these types of cases will be assigned the case. Thus, to make the cases to be analyzed for the RCG:91 more consistent with the assumption of random assignments, the cases that were assigned to these categories were removed from the analysis file, as described in a previous section.

Limiting the cases to be analyzed to those that were not assigned to specialized interviewers eliminates many of the most serious deviations from the theoretical, random assignment model. However, there were other non-random factors that might make the model inappropriate. For example, some interviewers only conducted interviews during the daytime hours. If graduates that could be reached during the daytime were systematically different from other sampled graduates (e.g., all were unemployed), then this could result in confounding the estimates of the interviewer effects with the characteristics of the cases and overestimating the correlation coefficient.

One way of accounting for these nonrandom factors is to explicitly include them in the model as **fixed effects**. In this case, fixed effects are attributes of the data collection process that are specific to the RCG:91 and are not considered a sample from a larger population. These effects can be included in the model as follows:

$$y_{kji} = \alpha_k + \beta_j + \tau_{kji}, \tag{4.8}$$

where the α term is a general fixed effect, and k is a subscript for the fixed effects. The new error term (τ) accounts for all the deviations from the fixed and random effects in the model.

The fixed effects included for the RCG:91 were

- Telephone center location (two Westat telephone centers);
- Month of interview (three values: June-July, August-October, and November-December);
- Time of interview (three values: before 5 PM, 5-8 PM, after 8 PM); and
- Time zone of interview (four values: Eastern, Central, Mountain, and Pacific).

The time of the interview and the time zone variables refer to the respondent's time, not that of the interviewer.

As noted earlier, the goal of this research was to estimate the interviewer contribution to the variance. The estimation of the significance of the fixed effects is not required, so model (4.8), which aggregates all the fixed effects, is appropriate for this purpose.

The model (4.8) is called a **mixed model**, because it involves both fixed (CATI site, etc.) and random (interviewer) effects. Statistical software that accommodates the estimation of the random component of the variance in this type of model exists, but no software was found that correctly accounts for the differential weights of the RCG:91. As a result, the subsampling of the cases to make the analysis file self-weighting, described earlier, was employed.

The VARCOMP SAS procedure was then used to implement the estimation of the random component of the error. A restricted maximum likelihood method of estimation of the parameters was used. See SAS (1989) for details on the procedure. Basically, the output of the procedure produces the variance component for the random interviewer effect and for the error term. The estimated correlation coefficient is the ratio of the interviewer component to the sum of the interviewer and error components.

Special Concerns for Dichotomous Variables

The error structure of dichotomous variables presents other concerns that must be addressed to ensure that the model provides an appropriate representation of the process. The two main considerations are the assumptions of the homogeneity of the variance and the normality of the effects. We begin with the homogeneity assumption, which is the more serious of the two concerns.

In the model, the variance of the response variable after accounting for the fixed effects is assumed to be constant across interviewers. When the response variable is a percentage, then this homogeneity assumption may not be satisfied because the variance of a percentage is a function of the percentage. If interviewer effects are present, then the percentage varies across interviewers, invalidating the assumption of constant variance. Stokes and Mulry (1987) and Stokes (1988) have discussed this problem in more detail.

The variance of a percentage is relatively constant for percentages that range between 20 and 80 percent (the variance goes from 16 percent to 25 percent in this range). The violation of the homogeneity assumption is most likely to occur for percentages less than 10 percent or greater than 90 percent.

Because of this concern, the questions selected for inclusion in this analysis were generally limited to those that had estimates in the general range of 20 to 80 percent. The restriction was imposed to avoid more complex estimation procedures required to properly handle extreme percents.

This restriction also helps alleviate the distributional assumptions associated with tests of significance and confidence intervals. These types of statements are based on the assumption that the response variables and the interviewer effects in model (4.8) are normally distributed. These inferences are generally robust to moderate deviations from the normality assumption. If extreme percentages are not included in the analysis, the robustness of these procedures should provide protection against invalid inferences.

FINDINGS

In this section, the results of applying the methods described above to selected questions from the RCG:91 are presented and discussed. The analysis file used for the computations was the self-weighting file described earlier. In all, 44 questions were selected for the analysis, including several different types of questions that might be expected to vary in terms of interviewer effect. All of the questions, except one, were treated as dichotomous variables by collapsing response categories, when required. Most of the estimated proportions ranged between 20 and 80 percent, although a few questions slightly outside this range were included.

The one question that was not dichotomous was SUMY106, which is a count of all of the yes responses to questions about the sources of financial support that the graduate received. The interviewer read 12 different sources of financial support and asked if the graduate used each of these. Thus, the SUMY106 variable could take on any value between 0 and 12.

General Comments

The estimated intra-interviewer correlation coefficients are shown in Table 4-1 for the selected questions from the RCG:91 (see Figure 4-1). The questions are listed in the order in which they appeared in the interview.

The most important and obvious finding is the small size of the intra-interviewer correlation across nearly all the questions examined. Two-thirds (30) of the 44 questions in the table have estimated intra-interviewer correlations of 0.005 or less. Only 4 of the questions have estimated correlations of 0.02 or greater. The mean of the estimated correlation coefficients shown in Table 4-1 is 0.008, and the standard deviation of these estimates is 0.015. The standard deviation is larger than might be expected due primarily to the inclusion of a few large estimates associated with question 51.

Table 4-1. Estimated intra-interviewer correlation for selected questions

Item		Sample size	Estimated percent	Intra-interviewer correlation[5]
Q10	Gradepoint average for undergrad level	8,655	85%	0.002
Q12	Has R attended since receiving degree?	8,761	34	0.001
Q15	Is respondent still enrolled?	2,992	62	0.005
Q17[1]	Type of school R was attending	2,851	16	0.006
Q18	Is this a public or private institution?	2,959	69	0.000
Q22	Was R attending full or part time?	2,990	51	0.003
Q23	Was R working for pay in reference week?	8,742	85	0.001
Q24	Was R looking for work in reference week?	1,340	34	0.100
Q25	Was R available to work in reference week?	1,334	40	0.000
Q33	Would R have wanted full-time job?	921	38	0.015
Q34[2]	What kind of employee was respondent?	7,355	69	0.001
Q35	Was business incorporated or not?	207	21	0.058
Q42	Was college degree required for main job?	7,326	63	0.008
Q43[1]	How close was major related to main job?	7,388	21	0.000
Q46	Was R looking for another job - Apr 22?	7,387	22	0.000
Q50	Is R eligible to teach at any level?	8,720	17	0.002
Q51PREK	Eligible prekindergarten	1,173	30	0.034
Q51KIND	Eligible kindergarten	1,173	61	0.007
Q511ST	Eligible first grade	1,173	71	0.008
Q512ND	Eligible second grade	1,173	71	0.010
Q513RD	Eligible third grade	1,173	71	0.009
Q514TH	Eligible fourth grade	1,173	70	0.000
Q515TH	Eligible fifth grade	1,173	71	0.001
Q516TH	Eligible sixth grade	1,173	77	0.000
Q517TH	Eligible seventh grade	1,173	77	0.000
Q518TH	Eligible eighth grade	1,173	77	0.003
Q519TH	Eligible ninth grade	1,173	64	0.000
Q5110TH	Eligible tenth grade	1,173	60	0.007
Q5111TH	Eligible eleventh grade	1,173	60	0.002
Q5112TH	Eligible twelfth grade	1,173	60	0.006
Q51UNGR	Eligible ungraded	1,173	22	0.060
Q51ALL	Eligible all	1,173	22	0.057
Q53	Does R have certificate to teach school?	8,754	16	0.000
Q80	Did R teach Special ED students?	837	79	0.000
Q83	Did R take Spec Ed courses for credit?	639	61	0.000
Q85	Was teaching assignment full/part time?	832	86	0.000
Q96[3]	What was R's marital status in April 91?	8,754	61	0.001
Q106A	Expenses paid by earnings	8,741	78	0.004
Q106B	Expenses paid by work study	8,740	18	0.004
Q106D	Expenses paid by parents	8,739	59	0.004
Q106D2	Expenses paid by parents contributions	8,730	56	0.004
Q106G	Expenses paid by loans	8,739	41	0.000
Q106H	Expenses paid by grants/scholarships	8,675	45	0.000
QSUMY106[4]	Expenses source	8,761	--	0.004

--Summary variable where estimated percent cannot be calculated.
[1]Categories 1 and 2 combined to form percent.
[2]Categories 2 through 5 combined to form percent.
[3]Categories 2 and 3 combined to form percent.
[4]SUMY106 is a count of the number of yes responses to 1,173 each of the parts of item 106.
[5]The intra-interviewer correlation is defined by equation (4.6) in the text.
SOURCE: U.S. Department of Education, National Center for Education Statistics, *1991 Recent College Graduates Survey*.

Figure 4-1. Estimated intra-interviewer correlation coefficient, by size of estimate

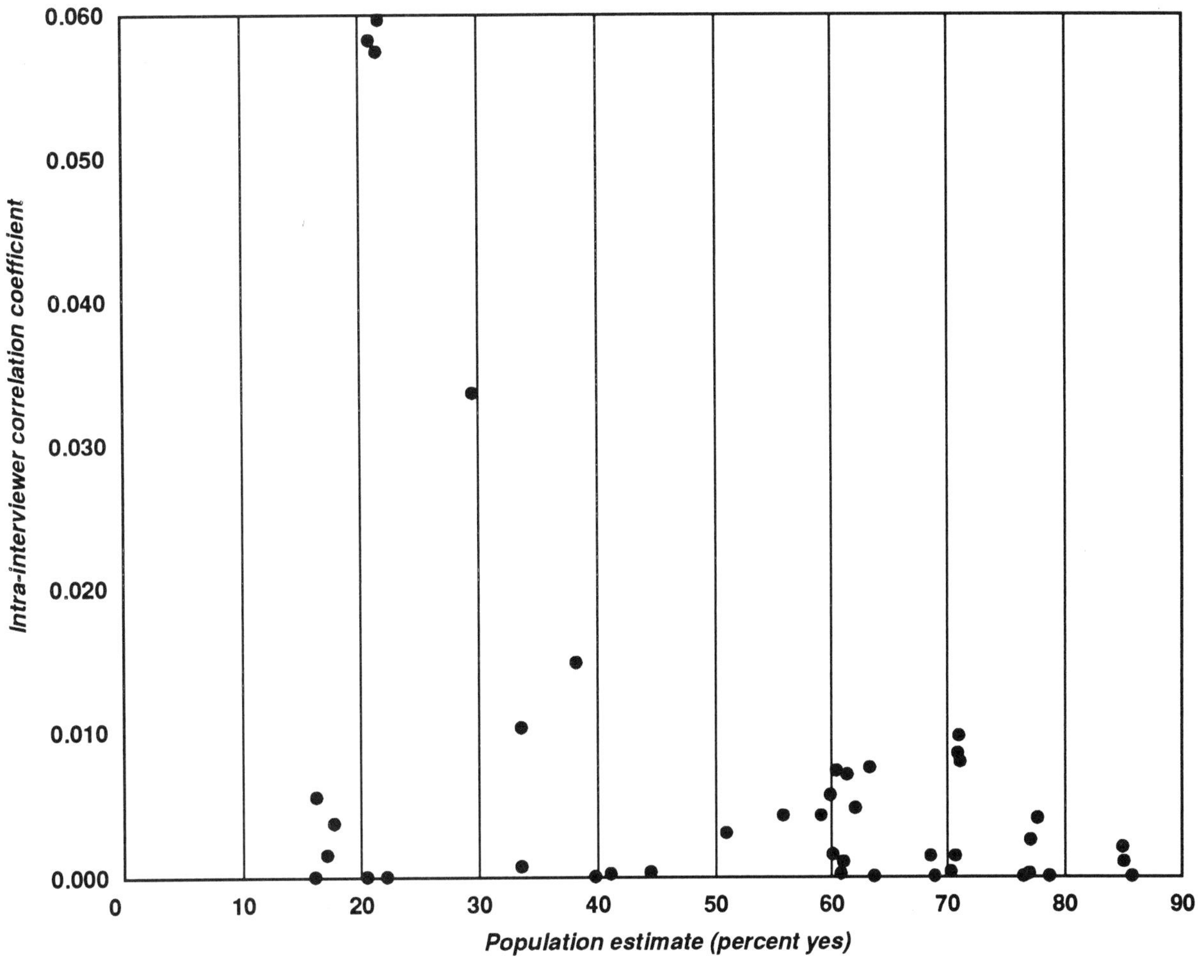

NOTE: The intra-interviewer correlation is defined by equation (4.6) in text.

The majority of the questions in the table (and in the interview in general) involved reading the question and getting a simple dichotomous response from the graduate, such as either yes or no, or full or part time. Of the 16 questions in the table of this type, only 1 (question 35) had an estimated correlation coefficient as large as 0.02. The interviewer effect for these items was very small, as might be expected given the nature of the questions.

Question 35 has an estimated intra-interviewer correlation of 0.058. This question was only asked for those graduates who reported they were self-employed in their own business. The question asked if the graduate's business was incorporated. Although there may have been some confusion about the definition of incorporated, the nature of the question is not one that is likely to exhibit larger interviewer effects.

It is more likely that the relatively large correlation for question 35 is due to the instability of the estimate. The estimate is based on just 207 observations (the next smallest sample size for any other estimate is 639). When the correlation was computed using the SAS GLM procedure without any fixed effects, the estimated correlation was only 0.002. This result suggests that the finding for this question is very unstable and may not be indicative of the high interviewer effects.

Multiple Response Questions

In addition to dichotomous response questions, there were some items that required the interviewer to read a question and a set of response options from which the respondent could choose. Questions of this type included question numbers 17, 34, 43, and 96. Question 10 asked the graduate for his/her grade point average, but the response categories were only read as a probe. The estimated correlations for all of these questions were also small, with none as large as 0.01.

Question 106 asked the respondent to respond either yes or no to 12 different sources of financial support. Again, the estimated correlations were small for all the various parts of this question as well as for the count variable (SUMY106) described above. All of the correlations were less than 0.01.

The largest correlations were estimated for question 51, in which graduates who said they were eligible to teach were asked what grades they were eligible to teach. This was an open-ended question for which up to 15 different grades could be recorded, beginning with prekindergarten, proceeding through 12th grade, including options for ungraded, and all grades. The interviewer was not intended to read the list of grades to the respondent.

The correlations for the grades from kindergarten through 12th grade were all relatively small, with none of these greater than 0.01. However, for prekindergarten, all grades, and ungraded, the correlations were larger, ranging from 0.03 to 0.06. Except for question 35, these are the only questions examined with large correlations.

The reason for the larger correlations for the three grades in question 51 can be inferred from the results and from observations on interviewer probing techniques. When an interviewer entered the code for all grades, the CATI system automatically coded all categories (except subject certified) as "yes," including prekindergarten and ungraded. Some interviewers routinely probed before using the all grades code to ensure that the respondent intended to include the prekindergarten, kindergarten, and ungraded categories. Other interviewers did not probe.

The intra-interviewer correlations are estimated based on the sample sizes shown in the table. Despite these relatively large sample sizes, the estimates of the correlations are subject to sampling variability. As Groves and Magilavy (1986) pointed out, the estimated correlations often have standard errors that are larger in size than the estimates themselves. This comment is likely to hold for the estimates from the RCG:91, although no estimates of the variability of the correlations were computed. Despite the poor precision for the estimates, even if the estimates were doubled, they would still be small for nearly all the questions.

Comparison to Other Studies

Groves and Magilavy (1986) summarized much of the literature on interviewer effects in telephone surveys. The studies they cite had mean estimated correlations ranging from 0.004 to 0.07, with most of the estimates less than 0.01. In many of the studies, the correlations were estimated from simple one-way ANOVA models and the designs were interpenetrating. Under these procedures, negative estimates of the correlations, which are nonnegative by definition, are not uncommon.

Groves and Magilavy also present interviewer effects from nine different studies at the centralized telephone facility of the Institute for Survey Research at the University of Michigan. They summarize the mean estimated correlations for this study as ranging from 0.002 to 0.016.

As noted earlier, the mean of the estimated correlation coefficients from the RCG:91 was 0.008. This average goes down to 0.006 if question 35 is excluded due to its small sample size. Both of these averages are consistent with the estimates from the other studies cited above. The design and estimation procedures used in the RCG:91 were slightly different from those other studies, but the general conclusions about the interviewer effects are essentially the same.

Impact on Variance of the Estimates

The estimated intra-interviewer correlations from the RCG:91 are a key ingredient to determine the impact of the interviewers on the reliability of the estimates from the RCG:91. The other factor required is the number of interviews conducted by the interviewers. Equation (4.7) demonstrates how these can be combined to estimate the inflation of the variance due to interviewers.

Table 4-2 shows the factor by which the standard error of the estimate goes up with different values of the correlation and the mean number of interviews per interviewer. This table was computed by taking the square root of the last factor in equation (4.7), since the standard error is used in inferences more often than the variance. When the correlation is small (0.005 or less), then the standard error only goes up a relatively small amount even for large interviewer loads. For larger correlations, the standard error can increase by 200 percent or more.

The mean interviewer load for RCG:91 was 115, which is the last row of the table. However, for estimates that were only asked for subsets of the population (e.g., teachers) the mean was smaller. For any particular question, the mean can be approximated by dividing the number of cases with responses by 106, the number of interviewers.

Example. For example, only about 12 percent of the respondents were teachers. The mean interviewer load for questions asked only of teachers is thus only about 12 percent of 115, or about 15. Thus, the increase in the standard error for these items could be found by looking at the first row of the table. If the particular question was a typical one (asking for specific responses), then the Table 4-1 findings suggest that the correlation is probably close to zero. The first two columns of the first row of Table 4-2 indicate that the standard error is probably an underestimate of about 2 to 5 percent due to the interviewer effect.

On the other hand, if the question was less structured, like the unusual responses to question 51, then the correlation might be as large as 0.05. The last columns of the same row of the table shows the standard error for the estimates of teachers is probably underestimated by 30 to 40 percent.

The results of Table 4-2 also show very clearly why the interviewer effects for master's degree graduates were not evaluated. With an average interviewer load of less than 15 even for those questions asked of all master's degree graduates, the impact of the interviewer effects on the standard errors of the estimates were likely to be small.

Table 4-2. Increase in the standard error of the estimate due to interviewer effects

Mean interviewer caseload	Intra-interviewer correlation						
	0.002	0.005	0.010	0.020	0.030	0.040	0.050
20	1.02	1.05	1.09	1.17	1.25	1.33	1.40
40	1.04	1.09	1.18	1.33	1.47	1.60	1.72
60	1.06	1.14	1.26	1.48	1.66	1.83	1.99
80	1.08	1.18	1.34	1.61	1.84	2.04	2.22
100	1.09	1.22	1.41	1.73	1.99	2.23	2.44
115	1.11	1.25	1.46	1.81	2.10	2.36	2.59

SOURCE: U.S. Department of Education, National Center for Education Statistics, *1991 Recent College Graduates Survey.*

IMPLICATIONS AND RECOMMENDATIONS ON INTERVIEWER EFFECTS

The findings presented above provide a mechanism for users to evaluate the probable impact of the interviewers on the standard errors of the estimates from the RCG:91. The average interviewer load can be computed directly by dividing the number of responses by 106 (the number of interviewers). If the question was included in Table 4-1, then the inflation in the standard error of the estimate can be computed directly.

In cases where the question was not studied, some subjective evaluation of the size of the correlation is required. However, the evaluation should be relatively simple. Most of the questions in the interview fall into types of questions similar to those studied in Table 4-1. Approximations of the correlations will provide adequate guidance for estimating the level of underestimation of the standard error.

Users can then exercise their own judgment about the need to modify their inferences to account for the underestimation. For example, if the underestimation of the standard error is less than 10 percent, many users may wish to ignore the factor in their analysis. However, if the underestimation is 100 percent, users may wish to double the estimated standard error when calculating confidence intervals or significance tests.

More complex analysis methods, such as regression analyses or differences between domain means, require additional research. Since many of these methods are based on estimates computed from small subsets of the sample, they may generally exhibit small interviewer effects.

These results also reaffirm the importance of structuring the interview in a consistent manner to avoid the undesirable impact of interviewer effects, especially in a centralized telephone operation in which the interviewer load is relatively high. Good questionnaire design and testing is extremely valuable in this regard. Given the findings presented above, the vast majority of the questions in the RCG:91 seem to satisfy these requirements.

An important step that could be undertaken to improve future RCG surveys is to review the entire questionnaire with the above findings in mind. Questions not included in the study could be evaluated from the prospective of interviewer effects. Clearly, some of the open-ended types of questions are the ones most likely to be problematic. Since the effect is dependent on how many respondents are asked the question, the effort could be concentrated on the questions that are asked of most or all of the sampled graduates.

Another step that could be taken to avoid interviewer effects in future studies is to increase the number of interviewers, thus reducing the mean interviewer load. While the number of interviewers is important in terms of measurement errors, it also has impacts on the schedule and the costs of the survey, especially training and supervision. All of these factors should be jointly considered in assessing the number of interviewers for future studies.

5. NONSAMPLING ERROR FROM MEASUREMENT ERROR: VALIDITY MEASURES

In the two previous chapters, measurement error was defined in terms of responses to repeated surveys conducted under the same general conditions. While this is a useful framework for understanding the nature and sources of important sources of measurement error, it is deficient in some areas. The most notable deficiency is in relating the survey estimates to the targeted population characteristics that might be constructed using different methods. For example, teacher certification by subject and level is a complex characteristic that can be estimated in several ways. The operational definition used in the RCG:91 might not conform with an administrative definitions used in each state. The repeated survey models may be of limited utility in estimating the equivalent of the administrative definition.

As discussed earlier, the difference between the survey estimate and the targeted population characteristic is due to both random variation and systematic bias. The random variation about the average from repeated surveys is estimated well using the models and framework described in the first chapters of this report. In addition, the systematic bias due to sources of error intrinsic to the survey, such as interviewers or coders, can also be handled with this approach. However, other sources of systematic bias are difficult to measure in this framework. The estimation of response bias from the reconciled reinterview data was noted as being a poor measure, largely because the reconciled reinterview could not be considered to be a much better estimate of the characteristic.

One way of avoiding this problem is to use an external data source that is not subject to the same sources of measurement error as the survey data. Comparing the survey estimates against this source can help identify differences that might be related to errors in either the survey or the external source. The resolution of the reasons for the differences is often difficult and may not lead to the identification of specific problems that can be remedied in the survey. Despite these problems, this type of benchmarking of the survey estimates against external data sources is valuable and has been done to some extent for the RCG:91.

This chapter uses a slightly different approach to help identify the accuracy of the results from the survey. The individual responses of the sampled graduates are compared to an external data source and used to help identify the nature and sources of the errors. Using models, these results are generalized to form estimates of the measurement error from the survey. By looking at the individual error terms in this fashion, it is possible to better understand the source of the errors and not just their global effects on specific aggregates.

Use of data from an external data source as the standard is predicated on the assumption that the data are free of error. Any deviations from the standard are considered as errors. This assumption must be questioned in actual practice because every method of data collection has its own sources of error. If the errors in the external source are small relative to the survey estimates, then the assumption may provide useful estimates of error. On the other hand, if the external source has substantial errors, then the error estimates using it as a standard may be severely overestimated.

A validity study was undertaken in the RCG:91 to examine the accuracy of teacher certification data reported by sampled graduates. Responses reported by graduates were compared to data provided by state teacher certification agencies, including both the type of certification and some attributes of the certification. Thus, the state teacher certification agencies provide the external data source for this evaluation.

Estimating measurement error by comparing the survey responses to external data opens up the possibility that the findings may differ from those of the reinterview analysis presented earlier. The reconciliation of the findings from the validity study and the reinterview will be one of the topics in the next chapter, along with the integration of all of the results of the assessments of measurement error.

Below, the purpose of the validity study and the procedures that were used to collect the data from the state certification agencies are presented. This is followed by a model for using these data along with the survey data to estimate measurement error, methods for estimating the error using the model, the application of these methods to the RCG:91 data, and a discussion of the implications of these findings for data users and designers of future studies.

PURPOSE OF THE VALIDITY STUDY

Some of the most important characteristics estimated from the series of RCG surveys are related to the number of new teachers graduated from higher education institutions. The number of graduates that are certified to teach and the kind of certification the graduates obtained are key variables for these estimates. If sampled graduates cannot adequately report these characteristics, the estimates from the series are of less use. Thus, a validity study was conducted to examine these issues for the RCG:91.

Self-reported data are often criticized for three different types of error, each of which could be applicable to the certification data in the RCG:91. These errors are discussed below:

- Deliberate or motivated errors are those in which the respondent adds or omits something in order to make a good impression. Potentially, teachers in certain situations might deliberately overreport their official certification. There may also be situations in which certification is pending or has been delayed for bureaucratic reasons and the teacher may report it as already achieved. Some teachers may overreport the areas in which they are certified if they are close to meeting the requirements.

- Lapses of memory are often a source of error in self-reported data, but this is not expected to be a significant source for the certification data. Teacher certification should be of relatively high salience. The graduates are mostly new teachers, and their certification is recent. However, for the 30 percent of the sample who were not teaching at the time of the interview, the certification information may be less salient. Those who were teaching in only one area may also have neglected to mention all fields in which they were certified.

- Question wording and problems with response categories are sources of errors in cases in which the respondent does not understand what is being asked or cannot fit the correct response into the choices given. This could be a significant source of error with certification data on the RCG:91 study. More response errors occur when questions are asked that have response units not normally used by those questioned to process the information (Marks and Mauldin, 1950). A wide variety of ways of expressing teacher certification exist throughout the country, yet the survey asked questions using response categories that attempted to be nationally applicable.

Design of the Validity Study

Because of the cost associated with contacting and gaining the cooperation of the state certification agencies, a two-stage design was used for the validity study. The design was not a probability sample, although selections were randomized. In the first stage, 10 states were selected from the 50 states and District of Columbia with probability proportionate to the number of sampled education majors who graduated within each state. No stratification was done due to the small number of states to be sampled.

The original state sample included Illinois, Virginia, and Hawaii. When these states were contacted to participate in the study, they indicated that they could not provide certification information without Social Security numbers. Since Social Security numbers were not available for the graduates, three other states (California, Ohio, and Tennessee) were substituted. The 10 states included in the final sample for the validity study were

Arkansas	Ohio	California	Indiana	Pennsylvania
Michigan	Utah	Florida	Texas	Tennessee

In the second stage, a simple random sample was selected within each sampled state from the graduates who reported that they were certified to teach in that state and had been interviewed on August 1, 1991, or later. For most states, a fixed sample size of 30 was selected. However, there were only 24 eligible graduates in Utah, so all were selected. To compensate, 46 graduates were selected in both Texas and Pennsylvania. In all, 326 were sampled for the study.

For confidentiality reasons, 30 graduates who were not included in the RCG:91 sample were included in the data requests sent to each state. These additional graduates were education majors who graduated from institutions in the sampled states and were selected from graduate lists from which the RCG:91 sample had been drawn. The data from these additional graduates were not included in the validity study analysis.

The survey form used to collect certification data from the state agencies used the same question wording and response categories that were used for the sampled graduates. A copy of the state survey form appears in Appendix E. The survey questions included whether the graduate was certified (question 53), the kind of certification (question 56), the grades certified to teach (question 54), and, the subjects certified to teach (question 59).

Data Collection

The appropriate person in each state agency was mailed a package of materials including a form for each graduate. The state agencies were asked to complete a survey form for each graduate and provide any written brochures or booklets explaining their certification procedures and requirements.

The information provided to the state for each graduate included the following (some items were not available for every graduate):

- Graduate's name and address at the time of interview;
- Alternate name information, such as maiden name or married name;
- Month and year of birth;
- Month and year of graduation; and
- Institution from which the respondent graduated.

Social Security numbers were not available for graduates.

During subsequent telephone contact with the state agencies, the agencies stated that their main concern was that searching by name was more time consuming and involved more complex procedures than searching by Social Security number. Nevertheless, all 10 states returned all of their survey forms for a 100 percent response rate. Nine of the states provided certification requirement materials.

All the state agencies indicated that without Social Security numbers they would need to use information such as birthdate to determine whether they had found the correct certification record. States may have used different definitions of what constituted a "match." For example, if the first and last name and birthdate were the same but the middle name was different, or the full name was the same but the birthdate was slightly different, were these counted as matches? In their study of survey matches to police records, Miller and Groves (1985) found that different definitions of a match resulted in a range of matches from 47 to 60 percent. The matching criteria for this study were left to the individual states.

Data Coding and Processing

A combination of manual and computer procedures was used to code, process, and analyze the certification data. The state survey forms were first manually edited and coded. Next, the state data were keyed and verified. The keyed data from the states were merged with the sampled graduates' RCG:91 data for analysis.

Manual editing and coding was used to provide the human judgment that could not be provided by computer. Miller and Groves (1985) examined procedures used to match survey responses to official records and found that "Machine matching procedures in some cases appear poor substitutes for human review that can simultaneously consider many variables and utilize any other information for matching that may be available...This selective supplementation of match criteria so easily performed by human review has no doubt led many past researchers to use human judgment to produce match decisions."[7]

This human judgment was needed primarily for two reasons. First, each state has its own categories, levels, and types of certification. How to report these individual categories on the standard survey form was subject to interpretation. Second, the data collection from graduates and states differed slightly. Graduates were read each category over the telephone and asked whether they were certified to teach that grade or subject. States were sent a form by mail and asked to circle the categories in which the graduate was certified. These different data collection methods may have caused some reporting differences.

Coding rules were established to handle situations in which the same certification had been reported differently by the graduate and the state. It should be noted that all editing and coding was done on the state certification forms; the information reported by the graduate on the RCG:91 survey was not changed. The coding scheme was designed so that the exact information reported by the state could be distinguished from codes assigned during processing. The specific coding instructions appear in Appendix F.

MEASUREMENT ERROR MODEL FOR VALIDITY DATA

In Chapter 3, model (3.11) was used to estimate response bias with reconciled reinterview data. A shortcoming of that application was the likelihood that the reconciled data were subject to much the same level of measurement error as the original interview data. This violated an important assumption of the model, making the estimates of bias suspect.

---

[7] Peter V. Miller and Robert M. Groves, "Matching Survey Responses to Official Records: An Exploration of Validity in Victimization Reporting," *Public Opinion Quarterly*, Vol. 49, No. 3, 1985.

The same model can be applied with the validity study data, with the hope that the data from the state agencies are error free or at least have smaller errors than the reconciled reinterview data. This would make the model assumptions more appropriate for the validity study.

Recall that in model (3.11) the measurement error arose only from the data collected in the original interview. The value from trial 2, which is from the state agencies in this case, is assumed to be the correct value. Thus, the state agency data are assumed to be unbiased and to have no response variance.

The response bias, as defined by equation (3.12), is a measure of how the estimate differs from the population value averaged over the entire population. In the next section, we will examine how the design of the validity study affects the ability to estimate this quantity.

The data from the state agencies can also be used to estimate a bound on the measurement error due to random errors (the SRV). The model is still that given in Chapter 3, but as with the response bias, the estimates that can be produced depend upon the validity study design. This is discussed further below.

Estimators

The estimators proposed for model (3.11) can be applied to the validity study and no new concepts are required, but the application depends on the design of the validity study. All of the validity data are categorical, so the general format for dichotomous data given in Chapter 3 is the natural way of presenting the results. Exhibit 5-1 is exactly as given there, except the rows now refer to the responses from the state agency reports rather than the reinterview results.

Exhibit 5-1. Interview by state agency responses

		Graduate reports		Total
		Number of cases with characteristic	Number of cases without characteristic	
State agency reports	Number of cases with characteristic	a	b	a + b
	Number of cases without characteristic	c	d	c + d
Total		a + c	b + d	n = a + b + c + d

The net difference rate, given in equation (3.14), was the estimator of response bias under the model proposed in Chapter 3. This estimator must be reconsidered within the context of the validity study for estimating the measurement error associated with certification.

The validity study design allows us to estimate the ratio of those who are confirmed as being certified by the state certification agencies to those who reported being certified on the survey, which is $a/(a+c)$. However, with the study design it is not possible to estimate the net effect of reporting errors on certification status, which is the survey estimate $(a+c)$ minus the state agency status estimate $(a+b)$ or the net difference rate as defined by equation (3.14).

The problem is that the responses were not validated for any respondents who said on the survey they were not certified. Since the respondents in the second column of the table were not sampled, it is not possible to directly estimate the b and d components. Therefore, the net difference rate as an estimator of the response bias cannot be estimated from the validity study data without making some assumptions.

There are two main reasons this approach was followed. First, it was assumed that b would be very small relative to $b+d$. To reliably estimate the proportion in the population who stated they were not certified but who in fact were certified would have required a substantial sample size. Since this was not the expected direction of the bias, the sample was allocated entirely to those who reported being certified.

Second, there were substantial operational difficulties that faced the states if those not certified were included in substantial numbers in the validity study. If the states found that a large proportion of the validity sample graduates were not certified, the chances for errors were likely to increase. The other operational concern was the number of graduates for which each state would be asked to search. For example, if the $b+d$ respondents were sampled, then the overall sample size would have at least doubled. Adding the nonsampled graduates included for confidentiality reasons, a total of 120 names would have been submitted to each state instead of 60. In fact, the sample would have been substantially larger than this to estimate the components reliably.

Therefore, the study was designed assuming b was very small, and the bias for the percentage of graduates certified must be estimated ignoring its contribution. Clearly an estimate ignoring this component is really an upper bound on the estimated bias, but this approximation may still be useful for determining the size of the bias. We study some of the consequences of the failure of this assumption later in this chapter and in the next chapter.

The **percent identically reported** (pir) is the percentage of the graduates who reported being certified that were confirmed as certified by the state agency. The percent identically reported can be written as:

$$pir = 100 \times \frac{a}{a+c} \qquad (5.1)$$

The percent not identically reported (100 - pir) is the net difference rate under the assumption that b is zero.

The estimators for the other characteristics collected from the state agencies can be formulated more generally, because they are based on the portion of the validity sample reported as being certified by both the graduate and the state agency. Thus, the graduate might have reported one kind of certificate and the state agency could have reported a different type of certificate, and all the cells of the survey by validity table could be filled. For the characteristics of these certified graduates, the more complete measures (the net difference rate and the gross difference rate) described in Chapter 3 were applied. The analysis of each of these questions (the kind of certificate, the grades certified to teach, and subjects certified to teach) is presented in terms of the gross and net difference rates.

The expected value of the net difference rate under model (3.11) is the response bias, as shown in Chapter 3. Under these assumptions, the expected value of the gross difference rate can be reduced from equation (3.6) to:

$$Eg = SRV_1 + \beta^2 \ , \tag{5.2}$$

when the characteristic is dichotomous, as it is for nearly all of the questions in the validity study. Thus, the gross difference rate estimates the simple response variance plus the response bias for these characteristics from the validity study.

The concerns raised above about using model (3.11) estimators for the confirmation of certification are an obvious effect of selecting a sample that was not representative of the whole population of college graduates. In other words, the response bias cannot be estimated for this characteristic because the sample was not representative of all graduates.

More subtle effects of the selection bias would be relevant for this estimate if graduates who reported they were not certified were included in the sample. For example, the fact that three of the states were substitutes raises questions about the theoretical framework for expressions such as (5.2) for the other characteristics of the validity study. The expectations of the estimators are not defined in the classical survey sampling framework. Even if no substitutes were used, the results of large sample theory would be of little use in a sample of 10 states of a universe of 50 states and the District of Columbia. Due to these limitations, no sampling weights were used in the analysis.

The implications of the possible selection bias should be considered in the use of the percent identically reported, and the net and gross difference estimates from the validity study. The use of formal randomized methods for selecting the graduates alleviates the most serious concerns about selection bias.

FINDINGS

The findings from the validity study for each certification item are discussed below, beginning with whether the graduate was certified, then looking at the kind of certification, the grades certified to teach, and the subjects certified to teach.

Certification to Teach

The first item examined is whether the graduate-reported certification was confirmed by the state certification agency. Table 5-1 displays the percentage of graduates with certification confirmed (*pir*), broken down by characteristics of the graduates as reported in the RCG:91 survey. The table also shows the percent not confirmed, which is an upper bound on the response bias estimated under the model (also see Figure 5-1).

Overall, 94.5 percent of the graduates in the validity sample had their certification confirmed. There was variability in the percent confirmed by state. Of the 10 states in the study, 5 states confirmed 100 percent of their sampled graduates as certified. Of the remaining 5 states, the confirmation rate varied from 80 to 96 percent.

Figure 5-1. Estimated percent of certificates not confirmed, by state

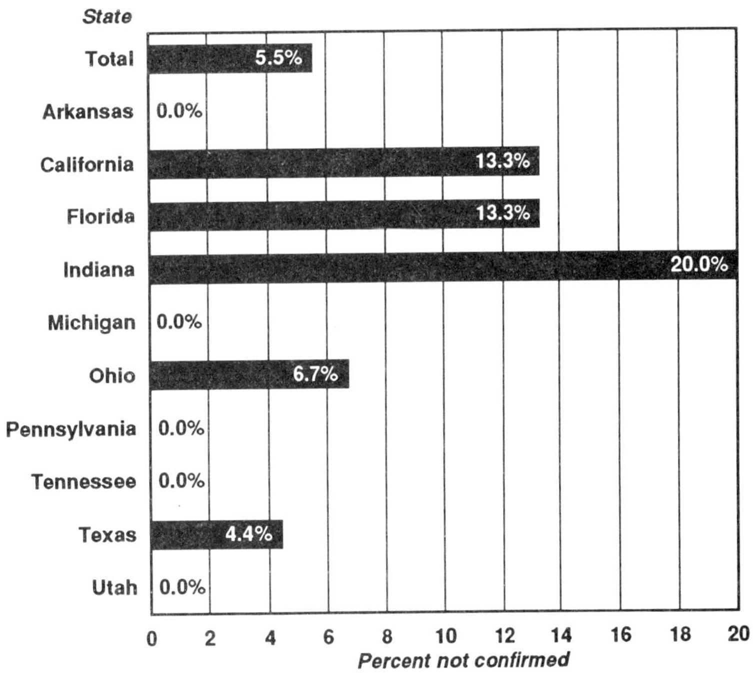

The estimates by state are important because the validity study data were collected by different state agencies, each of which could have contributed errors specific to the matching at the state level. The amount of flexibility each state had in its computer system to search for matches was different and may have contributed to the different confirmation rates by state. During telephone contacts with Westat staff, some states said that they had to work out special procedures with their computer staff for the search, since Social Security numbers were not available.

Each state was given the same information for searching their records, but the method for using the data was left to each state. For example, some states may have done their computer search by full name, while others may have searched by last name, then searched for an exact or close match by first and

Table 5-1. Percentage of graduates with certification confirmed and percentage not confirmed by state agencies, by graduate-reported characteristics

Category reported by graduate	Sample size	Percent confirmed	Percent not confirmed
Total	326	94.5%	5.5%
State of certification			
Arkansas	30	100.0	0.0
California	30	86.7	13.3
Florida	30	86.7	13.3
Indiana	30	80.0	20.0
Michigan	30	100.0	0.0
Ohio	30	93.3	6.7
Pennsylvania	46	100.0	0.0
Tennessee	30	100.0	0.0
Texas	46	95.6	4.4
Utah	24	100.0	0.0
Year certified			
Before 1989	36	86.1	13.9
1989	57	94.7	5.3
1990	188	97.9	2.1
1991	45	86.7	13.3
Kind of certificate			
Initial or provisional certificate leading to regular or standard certificate	133	97.7	2.3
Regular or standard	146	92.5	7.5
Alternative, emergency, or temporary certificate	34	88.2	11.8
Other	13	*	*
Certification level**			
Elementary	233	96.1	3.9
Secondary	93	90.3	9.7
Gender of graduate			
Male	74	86.5	13.5
Female	252	96.8	3.2
Teaching status			
Not teaching	98	88.8	11.2
Teaching contract or substitute teaching	228	96.9	3.1
Degree level			
Bachelor's	290	94.8	5.2
Master's	36	91.7	8.3

*Data suppressed because sample size was too small.
**Elementary includes graduates who reported certification in "any elementary fields" category in Q59. Secondary includes all others.
SOURCE: U.S. Department of Education, National Center for Education Statistics, *1991 Recent College Graduates Survey.*

middle name and birthdate. Large or small differences in names may have meant the difference between the state confirming or not confirming the certification. Thus, the variability in the percent confirmed by state may well have been related as much to the state's definition as to the graduate's error in reporting.

The confirmation rates also varied somewhat by the certification year reported by the graduate on the RCG:91 survey. In 1990, the confirmation rate was 98 percent and in 1989 was 95 percent. These 2 years contained three-fourths of the sampled graduates.

The variability in the percent confirmed by year suggests that another possible reason for differences may be associated with collecting the data at different points in time. The data from state certification agencies were collected from 3 to 9 months after the graduates were interviewed. The state survey form was designed so that if the graduate was not found to be currently certified, the state was asked to indicate whether the graduate was certified at any time during 1991. However, for 15 of the 18 graduates for whom certification was not confirmed by the state, the state could not determine whether a graduate had been certified at anytime during 1991. This points out another possible error associated with the state-reported data.

The lower rate for those who reported certification in 1991 may have been caused by graduates who confused being eligible with actually being certified, or who had applied but not yet obtained certification. The lower confirmation rate for graduates who reported certification before 1989 may have been partially caused by graduates who were certified at one time but did not renew their certification. Thus, the differences might also be indicative of a measurement problem in the survey.

The confirmation rates by the kind of certificate were highest for initial or provisional (98 percent) and regular or standard (93 percent). For the 10 percent of the sample that reported an alternative, emergency, or temporary certificate, the confirmation rate was 88 percent. It is possible that the lower confirmation rate for emergency and temporary certificates was related to the difference in graduate and state data collection time periods. An emergency or temporary certificate might be issued for a shorter time period than an initial or regular certificate. Therefore, a graduate might have held an emergency or temporary certificate at the time of the interview, but did not hold such certification when the state searched its records. Differences might also be due to different interpretations of what should be included as an alternative, emergency, or temporary certificate.

More than three-fourths of the validity sample were women, and women had a higher confirmation rate (97 percent confirmed) than men (87 percent confirmed). This would seem to indicate that name changes for women were not a major factor in the states' ability to locate the graduates' records. However, some of this may have been due to the data collection procedures performed specifically to avoid this problem, as discussed below.

When the completed certification forms were returned by the state, Westat staff separated the sampled cases that had not been confirmed. If any alternate or additional name information had been identified during the survey tracing or data collection (26 out of 46 unconfirmed graduates had such information), then that information was returned to the state agency. For the other 20 cases, the date and type of certification reported by the graduate were returned to the state agencies. Of the 26 cases with new name information, 25 were found by the states to have certification records. These findings indicate that the exact name used on the certification record was vital in determining whether the state found the graduate's certification record. Some of the cases where certification was not confirmed by the state might have been confirmed if additional name information had been available.

Of the 20 cases where the only new information was certification date and type, only 3 were found by the states to have certification records. While this percentage is small, it is interesting that any new matches were found, since the additional information provided should have made little difference to the search procedures. It is speculated that simply asking the states to search a second time resulted in more matches. These results again point out the assumption that the external data source is error free is very weak in actual implementation.

The analysis below examines the other characteristics collected in the validity sample. The analysis for these characteristics is restricted to the 308 sampled graduates that were confirmed as being certified by the state. Cases with missing responses to the specific question were dropped from the analysis.

Kind of Certification

Both the graduate and state were asked to choose one of the following categories for the kind of certification:

- Initial or provisional certificate leading to regular or standard certificate;
- Regular or standard;
- Alternative, emergency, or temporary certificate; and
- Other (specify).

Table 5-2 shows the percentage of cases in each category reported by the graduate and state. Two cases with missing data on kind of certificate were dropped from the analysis. As shown, most of the cases reported differently were those reported by the graduate as "initial or provisional" and by the state as "regular or standard" (24 percent of all the cases, and 56 percent of the cases reported differently by graduate and state were in this category). This difference may be partially due to the different time periods in which the data were collected. Graduates who had initial or provisional certificates at the time of the interview but obtained regular or standard certificates by the time of the state data collection would be in this category.

The gross and net difference rates for each state appear in Table 5-3, along with the aggregate over all 10 states. These rates are relatively large. The main reason for reporting differences in the kind of certificate appears to be different interpretations of the reporting categories. None of the 10 states included in the validity study use classifications exactly the same as those used on the survey. By looking at the classifications used in each state and the response patterns for that state, explanations for the reporting differences often emerge. A discussion of the state-by-state explanations is given in Appendix G.

Table 5-2. Percentage of all sampled cases by graduate-reported and state-reported kind of certification

Graduate-reported kind of certificate	State-reported kind of certificate				
	Initial, provisional	Regular, standard	Alternative, emergency, temporary	Other	Total
Initial or provisional	17%	24%	1%	--	42%
Regular or standard	10	33	1	--	44
Alternative, emergency, temporary	4	2	3	--	9
Other	--	--	--	4%	4
Total	31	59	5	4	100

-- Less than 0.5 percent.

NOTE: Details may not add to totals due to rounding.

SOURCE: U.S. Department of Education, National Center for Education Statistics, *1991 Recent College Graduates Survey.*

Table 5-3. Gross and net difference rates for kind of certificate from validity study, by state

State agency	Sample size	Gross difference rate	Net difference rate
Total	306	42.5	9.2
Arkansas	30	13.3	6.7
California	26	34.6	-11.5
Florida	26	46.2	-30.8
Indiana	24	20.8	20.8
Michigan	30	3.3	-3.3
Ohio	28	78.6	78.6
Pennsylvania	46	69.6	39.1
Tennessee	30	46.7	-46.7
Texas	43	44.2	44.2
Utah	23	52.2	-52.2

SOURCE: U.S. Department of Education, National Center for Education Statistics, *1991 Recent College Graduates Survey.*

Certification Grades

The graduates and state agencies were asked to report all grades that the graduate was certified to teach. The grade categories included prekindergarten, kindergarten, each grade from 1 through 12, ungraded, all grades, and subject certified. The ungraded category was meant to capture special education where certification may be given by ages rather than by grades. However, since no rules were set in the survey or on the state forms for use of this category, it is subject to interpretation. When interviewers entered the all grades category, the CATI system automatically coded all other categories "yes," except subject certified. As discussed in Chapter 4, interviewers differed in their use of the all grades category. The subject certified category was meant for graduates who were not certified by grade, but **only** by subject. Interviewers were instructed to always probe for grades before coding subject certified. If subject certified was coded, then no other grade category could be coded. However, some state agencies circled specific grades and subject certified to indicate that the graduate was certified both by grade and subject.

Table 5-4 gives the gross and net difference rates for each grade and the aggregate across all grades. These estimates are provided separately for all states and for all states except California. The rationale for excluding California follows.

Table 5-4. Gross and net difference rates for certification grade from validity study, by grade

Grade	Estimated percent of graduates certified in grade*	All sampled states		Excluding California	
		Gross difference rate	Net difference rate	Gross difference rate	Net difference rate
All grades combined	100%	8.9	-2.5	7.3	-0.4
Prekindergarten	27	12.5	-3.9	8.6	0.7
Kindergarten	61	11.1	3.9	11.1	5.4
First	71	4.9	-1.0	4.3	0.0
Second	71	5.6	-0.3	5.0	0.7
Third	71	5.2	-1.3	4.7	-0.4
Fourth	71	5.2	-0.7	4.7	0.4
Fifth	71	7.9	-2.0	7.5	-1.1
Sixth	76	10.8	-4.3	11.1	-3.9
Seventh	78	17.0	-3.9	16.8	-2.5
Eighth	78	16.7	-3.0	16.8	-1.8
Ninth	62	6.6	-2.0	4.3	0.7
Tenth	59	6.9	-2.3	4.3	0.7
Eleventh	58	6.9	-2.3	4.3	0.7
Twelfth	58	7.9	-3.3	5.4	-0.4
Ungraded	20	9.5	-5.6	4.3	0.0
All grades	20	9.5	-6.2	4.3	-0.7
Subject certified	1	6.2	-4.3	6.8	-4.7

*This is the estimated percent of all certified graduates who were certified in the grade, based on the weighted data obtained from the RCG:91 survey respondents (not just the respondents in the validity study).

NOTE: The sample size for all grades combined is 5,185 for the total sample and 4,743 for the states excluding California. The sample size for each grade individually is 305 (3 cases with missing grades were excluded) for all states and 279 excluding California.

SOURCE: U.S. Department of Education, National Center for Education Statistics, *1991 Recent College Graduates Survey.*

California has two types of teaching credentials, which differ somewhat from the other states in the sample. The authorization for teaching at each grade level for each certification type is as follows:

- Multiple Subject Teaching Credential. A teacher authorized for multiple subject instruction may be assigned, with his or her consent, to teach in any self-contained classroom (preschool, kindergarten, and grades 1 through 12, inclusive, or in classes organized primarily for adults); or to teach any subject in departmentalized classes to a given class or group of students in grade 8 and below, provided that the teacher has completed at least 12 semester units, or 6 upper division or graduate units of course work at an accredited institution in each subject to be taught.

- Single Subject Teaching Credential. A teacher authorized for single subject instruction may be assigned, with his or her consent, to teach any subject in his or her authorized fields at any grade level, preschool, kindergarten, and grades 1 through 12, inclusive, or in classes organized primarily for adults.

Thus, while multiple subject credential holders generally teach at the elementary level (which is usually taught in self-contained classrooms) and single subject credential holders generally teach at the secondary level (which is usually taught by subject), all California certified graduates can technically teach in any grade. The state, therefore, reported the grade level for all certified graduates as "all grades." The "all grades" category was translated as "yes" to every grade category except "subject certified" (including prekindergarten and ungraded). However, graduates often reported certification in fewer grades. As a result, the difference rates are lower when California is excluded (as shown in Figure 5-2). The gross difference rates for 9th through 12th grade are also lower when California is excluded. The most likely reason for this is the number of California multiple subject credential holders who reported that they were certified only in the elementary grades through grade 8.

For this reason, gross and net difference rates for the certification grade are examined for the 280 graduates outside of the California sample. The gross difference rate was 7.3 percent and the net difference rate only -0.4 percent. These estimates indicate relatively low measurement errors for these estimates.

The difference rates are highest in the transitional grades between elementary, middle school, junior high, and high school (grades prekindergarten, kindergarten, sixth, seventh, and eighth). The main reason for this was probably related to certification for different groups of grades. Elementary certification was sometimes reported through sixth grade and sometimes reported through eighth grade. Similarly, kindergarten might have been considered included in the elementary certification.

Figure 5-2. Estimated gross difference rates, by grade certified to teach

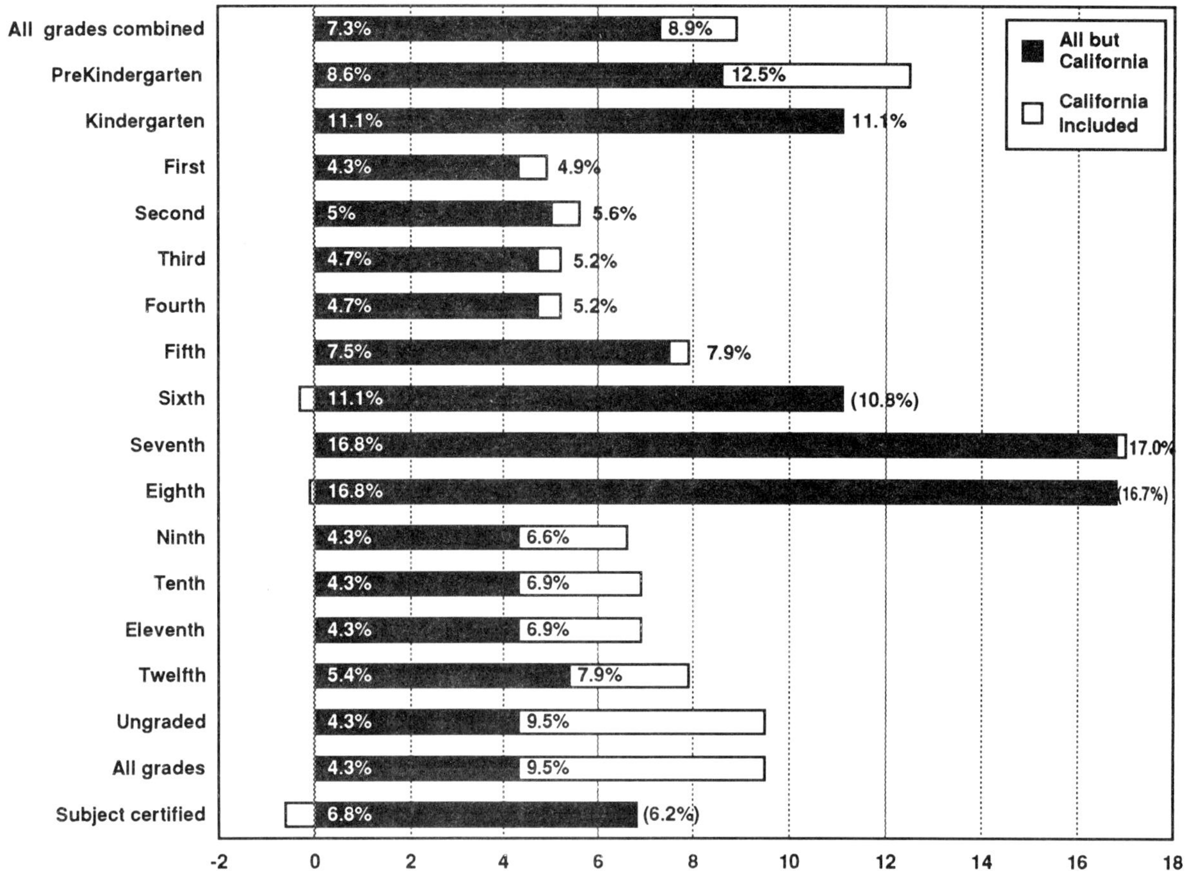

Note: Number in parentheses ( ) indicate that the inclusion of California resulted in decline in gross difference rate for these few grades.

Certification Subject Fields

The last RCG:91 survey question included in the validity study was subject fields certified to teach. Table 5-5 gives the gross and net difference rates for all subjects combined and for each of the subjects. These estimates are broken down as coded and uncoded following the procedures described below, which were designed to handle special problems with elementary and special education certificates.

The response category "any elementary fields, general or specialized" was intended to include any subject at the elementary level. An elementary certified graduate was expected to report a specific subject only if it was an additional certification beyond the general certification for elementary. In practice, many graduates said yes to each of the subject fields included in their elementary certification, rather than only those in addition to elementary. However, most of the state forms were completed according to the intent for elementary certification, i.e., the specific subject fields were circled only if the graduate had an additional certification in that field.

Table 5-5. Gross and net difference rates for subject field from validity study, by field

Subject	Estimated percent of graduates certified in subject*	Coded responses		Uncoded responses	
		Gross difference rate	Net difference rate	Gross difference rate	Net difference rate
All subjects combined	100%	1.8	1.2	10.4	9.6
Any elementary fields, general or specialized	68	2.0	1.3	11.6	10.9
Art/fine arts/performing arts	16	2.6	2.0	16.7	16.1
Basic skills and remedial education	25	3.3	3.3	29.5	29.5
Bilingual education	3	1.3	1.3	3.0	3.0
Biological or life sciences	19	2.3	1.6	18.0	17.4
Business	6	1.0	1.0	2.3	2.3
Computer science	8	0.7	0.7	9.5	9.5
English language arts	35	2.3	1.6	29.5	28.9
English as a second language	4	1.3	0.7	4.3	3.6
Foreign languages	4	1.3	-0.7	1.6	0.3
Gifted/talented	9	2.0	1.3	9.2	9.2
Health	21	3.3	2.6	21.0	20.3
Home economics	4	1.0	0.3	3.0	2.3
Industrial arts, trade, and industry	3	0.7	0.0	1.6	1.0
Mathematics	28	1.6	1.6	25.2	25.2
Music	12	0.7	0.7	11.5	11.5
Any physical sciences, general or specialized					
General sciences (no specialized area)	18	3.0	0.3	19.0	16.4
Chemistry	6	0.7	0.7	4.3	4.3
Geology/earth science	11	2.0	1.3	11.1	10.5
Physics	5	0.3	0.3	3.6	3.6
Other physical sciences	7	0.7	0.7	6.6	6.6
Physical education	16	2.6	2.0	13.1	13.1
Pre-elementary education	18	2.3	0.3	13.1	11.8
Reading	28	3.0	3.0	29.5	29.5
Religion/philosophy	5	1.6	1.6	5.6	5.6
Social science/social studies	32	3.6	1.6	25.6	23.6
Any special education field					
Mentally retarded	10	1.6	1.6	5.9	4.6
Hearing impaired, deaf	4	0.7	0.7	2.0	2.0
Seriously emotionally disturbed	8	1.3	1.3	3.0	1.0
Speech impaired	5	1.0	1.0	2.0	1.3
Specific learning disability	10	2.0	1.3	3.9	3.3
General certificate (no specific condition)	8	1.6	1.6	5.9	4.6
Other special education	5	1.3	1.3	4.6	2.6
Vocational education, other than business, home economics, or industrial arts	3	1.0	1.0	2.6	2.6
Other fields	8	4.3	-0.3	6.9	-1.6

*This is the estimated percent of all certified graduates who were certified in the subject, based on the weighted data obtained from the RCG:91 survey respondents (not just the respondents in the validity study).

NOTE: The sample size for all subjects combined is 10,675, and the sample size for each subject individually is 305.

SOURCE: U.S. Department of Education, National Center for Education Statistics, *1991 Recent College Graduates Survey.*

A second problem that occurred with elementary certification involved graduates certified to teach elementary grades but only in a specialized subject (such as physical education, art, music, reading). On the survey, some of these graduates reported "yes" to "any elementary fields," since they were certified on the elementary level. However, most of the states did not consider this to be elementary certification.

For graduates certified in special education, there were different interpretations of how to fit the certification into the survey categories. For example, graduates certified to teach "mildly handicapped K-12," frequently answered "yes" to the specific handicapping conditions that the certificate covered (such as mentally retarded and specific learning disability). However, the state often chose the category "general certificate, no specific condition," if no specific condition was named in the certificate. The different interpretations may have been increased by asking the graduates whether they were certified in each of the specific special education fields.

To address these problems, responses were coded so that only real differences in certification would appear in the difference rates. The coded responses can be used to examine the graduate's knowledge of their certification subjects outside of the reporting problems. For the coded responses, the gross and net difference rates were almost all below 4 percent, indicating that there was close agreement between the states and the graduates on certification fields. The net difference rates were only slightly lower than the gross difference rates, and almost all net difference rates were positive. This indicates that most of the small differences that existed were due to graduates overestimating their certification fields compared to the state-reported data.

The difference rates calculated using the uncoded responses are more pertinent to users who do not have the advantage of this postsurvey reconciliation. From this perspective, the uncoded responses are more indicative of the measurement errors in the survey estimates and are the focus of the findings.

The net difference rates were relatively large and almost always positive, indicating the tendency of the graduates to overstate the subjects they were certified to teach. The gross difference rates are only slightly larger than the net difference rates, showing the measurement errors were primarily from the response bias (see equation (5.2)). These errors were caused by graduates certified in elementary who also reported the subject fields included in their elementary certification. The subjects with especially large differences are those that are normally included in elementary certificates. This includes science, English, mathematics, reading, and social studies, as well as the specialized subjects of art, music, and health. In addition, many elementary teachers interpreted the "basic skills" category to mean the basic skills taught to all elementary students, which are included in their elementary certificates.

The problem with the special education fields appears to be smaller than expected, given the problems noted above. This is because the majority of graduates in the validity study sample were not certified in special education. The percentage of graduates certified in special education fields ranged from 4 percent to 10 percent, and the gross and net difference rates ranged from 2 percent to 6 percent in these fields.

IMPLICATIONS AND RECOMMENDATIONS FROM THE VALIDITY STUDY

The validity study data provide another method of examining the impact of measurement errors on the estimates from the RCG:91. The validity study data were limited to estimates related to certification.

For the percentage of graduates certified to teach, the validity study provides an upper bound on the response bias. The findings showed that the response bias due to overreporting being certified to teach was less than 5.5 percent. The examination of these results by state and year suggest that errors in matching the graduate at the state level may have resulted in overstating the response bias. Thus, it is very likely that the net response bias for this item is considerably less than 5 percent.

Based on these results, the estimates from the RCG:91 should be considered to be fairly accurate for most purposes. No specific estimation procedures or adjustments of the RCG:91 estimates of the number and percent certified to teach are recommended.

The model assumptions are more closely met for estimating the kind of certificate, the grades certified to teach, and the subjects certified to teach. For the kind of certificate, the response bias and variance appeared to be substantial. However, some of these errors may have been the result of the different time periods of data collection rather than measurement errors in the survey estimates.

Teachers with regular or standard certificates appear to be estimated well using the current RCG questions and procedures, but nonstandard kinds of certificates are probably subject to more measurement error. Because of the time delay in reporting, users are not recommended to make any adjustments in their analysis, based on these findings. However, some investigation of the handling of nonstandard certificates is warranted for future surveys.

The measurement error for estimates of the grades certified to teach were relatively small. The errors were somewhat larger for grades that are at the transition for one grade level to the next (elementary to secondary), and these errors might be addressed by some changes in the way the question is formulated. Again, users are not advised to alter their analytic methods due to these findings.

The response biases for the subjects certified to teach were relatively large, especially for subjects that are normally included in a general elementary certificate. Users of the results from this question should be concerned about the reliability of these estimates. Rather than using coarse adjustments, users should consider producing estimates that avoid the overestimates to the extent possible. For example, the estimates of subjects certified to teach could be restricted to the subset of graduates who reported they were not certified in the category "any elementary fields," since the problem is largely associated with the elementary grade teachers. At a minimum, estimates of the subjects that have the largest estimated response biases should be noted in any analysis and the reasons for the overestimation should be discussed. The same procedure should be followed for the eligible to teach data.

Clearly, the RCG:91 instrument needs to be revised to account for this problem before the survey is conducted again. Even simple changes, such as not reading each of the subjects, could have a major impact in reducing the bias for the estimates for these characteristics. Suggested questionnaire revisions appear in Appendix H.

6. SYNTHESIZING MEASUREMENT ERRORS FROM DIFFERENT SOURCES

Three major sources of nonsampling error in the RCG:91 were examined in the previous chapters of this report. These were errors due to nonresponse, random errors due to measurement problems, and systematic errors due to interviewers. The validity study provided a different way of looking at these errors, but only for a few estimates related to teacher certification. As these errors were studied, a discussion of the potential consequences for users of the RCG:91 data and suggested areas for improvement in the survey process were included.

In this chapter, these disparate results are organized into a more comprehensive overview of the survey errors in the RCG:91. The findings from the reinterview study and the validity study are first reviewed to assess the consistency between the two. This step is important because it helps guide the development of a more complete measure of nonsampling error.

After this evaluation, the development of a more integrated model including both nonresponse and measurement error components is discussed. Several approaches to combining different errors are considered and the limitations of each method are presented. A model is adopted for subsequent analysis, but even this approach is not without problems.

The final section presents some overall recommendations for data users and designers of future RCG surveys. These recommendations are a sysnthesis of earlier discussions.

COMPARING REINTERVIEW AND VALIDITY STUDY ESTIMATES

The validity and reinterview samples were selected independently, and the overlap in terms of graduates included in both is too small to support any direct comparisons. However, two of the questions covered in the validity study were also in the reinterview study, and the estimates of measurement error for these can be compared. The questions included in both studies were question 53, whether the graduate was certified to teach, and question 59, the subjects certified to teach. We begin with the certified to teach question.

The response bias of the estimate of the number of graduates certified to teach was computed in Chapter 3, using the net difference rate based on the weighted original and reconciled reinterview data. The survey by reinterview (reconciled) table for this question is given in terms of percentages (Table 6-1).

The net difference rate estimated from this table for the certified to teach statistic is 0.1 percent, with an estimated standard error of 0.3 percent. Thus, based on the reconciled reinterview data, the net bias in the estimated number of graduates who are certified to teach is small and not significantly different from zero.

Table 6-1. Percent certified to teach, interview-reinterview results

		Original interview		Total
		Certified to teach	Not certified to teach	
Reinterview (reconciled)	Certified to teach	14.51%	0.70%	15.21%
	Not certified to teach	0.56	84.23	84.79
Total		15.07	84.93	100.00

The data in the first column of Table 6-1 can also be used to estimate the percent identically reported for those graduates who originally reported being certified to teach. Using equation (5.1), the percent identically reported is 96.3 percent ($\frac{100 \times 14.51}{15.07} = 96.3\%$). Thus, an upper bound on response bias based on the reinterview data would be estimated by the percent not identically reported, or 3.7 percent.

Earlier in Chapter 5, we computed an estimate of the upper bound on the response bias using the percent not identically reported based on the validity study data. Since the reinterview only included bachelor's recipients, the comparable estimate from the validity study based only on bachelor's recipients is 5.2 percent (275 of the 290 bachelor's recipients in the validity study were identified as certified to teach by state certification agencies).

The estimated upper bound on the response bias from the reinterview study of 3.7 percent is within sampling error of the estimate of 5.2 percent from the validity study. These results are not inconsistent with each other; however, there are other factors that are related to the consistency of the two studies.

As discussed in Chapter 5, some errors in the matching with state agency records undoubtedly resulted in an overestimate of the bound on the response bias. In particular, the process of matching without Social Security numbers and the time period difference between the survey and validity study worked to artificially inflate the response bias bound. Two manifestations of these problems were the additional matches when the sampled graduates were sent back to the agencies a second time and the unexpected variability in the number not confirmed as certified to teach by state.[8]

The basic problem is with the validity study design assumption that the percent of the graduates who reported that they were not certified but actually were certified was sufficiently small that the bias (or at least a bound on the bias) could be estimated without sampling from this group. The assumption was not consistent with some of the other observations from the

[8] The variability in the number not certified varies more by state than would be expected under a simple model. Assume the errors in reporting certification are distributed as a poisson random variable, independent of state. Using the mean rate of 1.8 errors per sample and the data in Table 5-1, the observed distribution by state has more states with all confirmed and a few states with many more not confirmed than would be expected.

validity study and, perhaps even more significantly, with data from the reinterview. The estimated numbers in the off-diagonal cells of Table 6-1 are nearly equal, which demonstrates that the assumption is not supported by the reinterview findings.

These results suggest that the validity study is grossly overestimating the response bias for the number certified to teach. As noted above, this overestimate occurs because the assumption that all graduates who reported they were not certified to teach did so without error does not hold. Furthermore, the assumption that the verification with the state agencies was done without error is also not true.

Another possible explanation for the inconsistency is that the reinterview study does not measure response bias. This would happen if the reconciliation process did not significantly improve the accuracy of the respondents' reports for the certification question. Although possible, the finding that the percent identically reported from the reinterview data is consistent with the validity study findings for those graduates who originally reported being certified contradicts this explanation.

The evidence from these studies leads us to suspect that the reinterview estimates of bias are better approximations of response bias than the estimates from the validity study for the certified to teach question. The assumptions for the reinterview model are more supported by the data and seem more reasonable.

Comparing Error Estimates for Subjects Certified to Teach

The other question included in both the validity study and the reinterview study was on subjects certified to teach. For this item, the reinterview results were not reconciled, so no appropriate estimate of response bias can be obtained from the reinterview data. Thus, attention is restricted to comparing estimates of the simple response variance from the two studies to determine if these are consistent.

In the validity study, the gross difference rate for the subjects certified to teach was estimated to be 10.4 percent and the net difference rate was 9.1 percent. The gross difference rate under the validity study model includes both the simple response variance and the response bias (see equation (5.2)), while the net difference rate (computed from the unreconciled data) estimates the response bias. Since the net difference rate is large compared to the gross difference rate, the estimate of the simple response variance based on the validity study results is small for this question.

From the reinterview data, the gross difference rate for this question was estimated as 8.6 percent. Under the reinterview model (3.2), the gross difference rate divided by two is an unbiased estimate of the simple response variance. Therefore, the estimate of the SRV from the reinterview data is 4.3 percent.

The estimates from the validity study and the reinterview study both show that the SRV is small for the subjects certified to teach question. While the estimate of the response bias from the validity study is relatively large, this finding cannot be compared to findings from the reinterview study, since this question was not reconciled.

MORE COMPLETE MODELS OF NONSAMPLING ERRORS

The sources of nonsampling error investigated in this study were the three that were thought to have the greatest potential for distorting inferences from the RCG:91 and those with at least some resources devoted to their evaluation. In addition to those studied, other sources of nonsampling error could result in biases and additional variation in the estimates. Since these other sources of error were not included in the evaluation efforts, they are not discussed below.

The methods used to evaluate the errors in the RCG:91 estimates were based on specific models and assumptions about the distributions of the parameters of the models. Whenever these types of models are used, the robustness of the models and the appropriateness of the assumptions should be questioned. If the models or assumptions are inadequate, then the estimates of the nonsampling error derived from them may be misleading. In each chapter, the assumptions of the models were explicitly stated and evidence from the studies were used to investigate the reasonableness of the models and the assumptions. Unfortunately, this type of checking was not always possible.

Efforts to extend the models in the previous chapters to jointly account for graduate nonresponse error, random measurement error, and interviewer-related measurement error have been attempted. The work of Bailar and Biemer (1984) is consistent with the nonsampling error models presented for the RCG:91.

They begin with the model:

$$y_{ji}=\mu_{ji}+e_{ji}, \tag{6.1}$$

where the terms are defined as before, with i designating the sampled graduate and j the interviewer. The error term is now defined so that both nonresponse and measurement error can be explicitly included:

$$e_{ji}=(1-\delta_{ji})e_{ji}^{(1)}+\delta_{ji}e_{ji}^{(2)}, \tag{6.2}$$

where

δ_{ji} = 1 if the graduate responds

= 0 if the graduate does not respond

and

$e_{ji}^{(1)}$ = the error from nonresponse

$e_{ji}^{(2)}$ = the error from measurement

In this extended model, the error term included depends on whether or not the sampled graduate responds to the interview. If the graduate responds, then the error term is like the ones studied in Chapter 3 and 4 for errors associated with measurement, including the contribution of interviewers. If the graduate does not respond, then the error term is an imputation error (or the difference in the estimates after the weighting adjustments), such as that covered in Chapter 2. The error terms may have nonzero means, variances, and covariances. Furthermore, the response indicator, δ, is a random variable and has its own distribution and can be correlated with the measurement and nonresponse errors.

Bailar and Biemer (1984) present this model and make some simplifying assumptions about the distributions in order to arrive at some general statements about nonsampling errors. They stop short of computing estimates of the error under the model. This is due in part to a lack of information on the terms required and also because the assumptions required to make the model tractable are likely to be violated. The same problems they encountered prohibit using this model to integrate the errors from nonresponse and measurement sources in the RCG:91.

One of the important hurdles that makes it difficult to model these different error sources together is the interaction in the errors. Simple additive models are clearly inappropriate for the RCG:91. For example, a model that posits that the mean square error of the estimate can be represented by:

$$MSE(\hat{y}) = (SV + SRV) \times (1 + (m-1)\rho) + B_{nr} + B_M \;, \tag{6.3}$$

where the first term includes the variable errors due to measurement error and the systematic error due to interviewers and the last two terms are the bias due to nonresponse and measurement bias, respectively.

This model assumes that the nonresponse error and the measurement error are additive. This is not supported by other research findings. For example, almost all studies of nonrespondents, including the analysis in Chapter 2, show that nonrespondents have different characteristics than respondents. Measurement errors are also different for different groups of respondents. It is very likely that there are correlations between measurement errors and the probabilities of the graduates responding.

The lack of a term for the interaction between nonresponse and measurement errors is one of the most significant deficiencies of the model suggested by (6.1) and (6.2). Such a term could be added to the model, but this introduces the problems that Bailar and Biemer faced with their model. With the interaction term, the model becomes intractable even when some simplifying assumptions are made.

A different approach to incorporating both nonresponse and measurement errors was attempted by Anderson et al. (1979). They studied three forms of bias (nonresponse, field, and processing) using the general approach suggested by Kish (1965). In concentrating on bias, they assumed that the standard error of the estimate included the most important sources of random or variable error.

Their method of estimating bias was to compare the survey estimates to data from external sources and attribute differences between the two to bias in the survey. This required the assumption that the external source was a standard that had either no measurement error or very small errors relative to the survey.

This approach could be applied to the RCG:91, but most of the important estimates from the survey cannot be compared to external sources because these sources do not exist. We also saw that our attempts to do this type of comparison for a validity study of questions related to certification to teach were limited because of errors in the matching of graduates to the state agency records. Furthermore, this approach does not include the interviewer contribution to the nonsampling errors, which was found to be an important source of nonsampling errors for some characteristics in the RCG:91.

Given the difficulty in finding an appropriate model for both nonresponse and measurement error that yields estimates from available data, a less structured method of assessing the joint effects of nonsampling errors for the RCG:91 is presented in the next section. The findings from the earlier chapters are first discussed in general terms and then recommendations for making statistical statements from the survey estimates in the presence of these errors are presented. While this approach does not eliminate the problems discussed above, we hope that results will be interpreted critically.

REVIEW OF FINDINGS

Nonresponse bias in the RCG:91 was most likely to arise because not all the sampled graduates were interviewed. The increase in the variance of the estimates due to the nonresponse adjustments could be important, but the estimated sampling errors contained a contribution for this inflation. Because of the high response rate for most items, missing values from other responding graduates were not found to be significant problems for most characteristics. Thus, the potential bias in the estimates is the primary unaccounted effect of the nonsampling error due to nonresponse.

In Chapter 2, the impact of the nonresponse bias was found to be most significant for estimates based on large sample sizes, especially when the characteristic being estimated was correlated with the response rate. The adjustment procedures used, including the nonresponse and poststratification adjustments, were found to reduce the bias for many estimates.

Using the models of measurement error from Chapter 3, it was shown that the errors that were not systematic biases were already included in the sampling errors computed from the survey. Response bias was also studied by reconciling the responses from the original interview with those given in the reinterview. Generally, the response biases estimated using this procedure were small and not statistically significant.

In Chapter 4, the measurement error model was extended further to include the systematic errors associated with interviewers. The results of this study showed that the intra-interviewer correlations were very small, but the effects on the standard errors of the estimates could still be significant because of the large interviewer caseload. The effects were expected to be largest for characteristics that were asked of all or most sampled graduates, since these were the questions with the largest caseloads.

From these general conclusions and from the specific estimates of nonsampling errors presented earlier, it is possible to speculate on the nonsampling errors for some of the characteristics from the RCG:91 and on methods of analysis to account for these errors. This is done for several examples below.

Example: Working for Pay

Question 23 asked if the respondent was working for pay in the reference week. The unadjusted estimate from the survey was that 84 percent of the bachelor's recipients working for pay in the reference week. Using the standard methods of analysis, the standard error of this estimate is estimated to be 0.29 percent and the 95 percent confidence interval is 83.4 percent to 84.6 percent.

Consider now the adjustments that might be considered to account for the nonsampling errors for this estimate. The net difference rate was estimated from the reconciled reinterview as 0.7 percent, but this estimate was not significantly different from zero. If we assume that the nonresponse is correlated with the employment question and is largely unaffected by the nonresponse adjustments, we might expect the nonresponse bias to be between 0.3 and 1.0 percent for the estimate. We might use 1 percent as an upper bound on the bias due to both of these sources. Therefore, a bias adjusted estimate would be 83 percent (survey estimate minus the estimated bias).

The simple response variance is already included in the estimate of the standard error of the estimate, so the only adjustment is for the systematic error associated with the interviewers. Assuming that the intra-interviewer correlation is small for this question (as it is for most questions with simple yes/no response categories), the estimated multiplier for the standard error is 1.11 (the entry in first column of the last row of Table 4-2). Thus, the adjusted standard error for the estimate is 0.32 percent (1.11 times 0.29).

Incorporating adjustments for both the bias and the standard error, the adjusted estimate is 83 percent, with a 95 percent confidence interval from 82.3 percent to 83.7 percent. This assumes that the bias adjustment does not impact on the variance of the estimate. A further discussion of this issue is postponed until later. The difference between the adjusted and unadjusted confidence intervals is small for nearly all substantive uses. Furthermore, the bias adjustment is poorly estimated from the data at hand and is more an upper bound than a point estimate.

A better alternative, suggested in Chapter 2, is to use a more conservative approach for forming the confidence intervals to guard against the effects of possible bias. The adjustment for the interviewer effects is still warranted, even though it is generally small. Using this method, the confidence interval is computed by multiplying the adjusted standard error of the estimate by 3 instead of 2, and the estimated confidence interval extends from 83 to 85 percent.

Example: Working for Pay Subdomain Estimates

While some interest lies in the estimates for all graduates, many of the most important substantive findings are those that compare the estimates from one domain with those from another. For example, interest often centers on estimates such as the difference between the percent of males and females who are working for pay. Since the implications of the nonsampling errors for these types of estimates are different, this type of estimate is explored below.

The estimated difference between the percent of males and females working for pay can be written as:

$$\hat{d} = \hat{p}_{males} - \hat{p}_{females} , \tag{6.4}$$

where the estimates are based on the unadjusted survey estimates. If an adjustment for nonresponse and measurement error bias were to be included, the adjusted estimate could be written as:

$$\hat{d} = \hat{p}_{males} - \hat{p}_{females} + \{ \hat{b}_{males} - \hat{b}_{females} \}, \tag{6.5}$$

where the terms in brackets are the bias adjustments for the estimated percent of males and females working for pay, respectively.

While no estimates were computed for the nonresponse and measurement bias adjustments for males and females separately, the net bias (the term in brackets) is probably smaller than the bias for each of the estimates individually. Thus, the bias adjustment should probably be less for the estimated difference than for the estimate for all graduates.

While this result is somewhat comforting, the fact remains that the bias relative to the estimate may be larger for the estimated difference than for the overall estimate. The bias depends on the difference in the percent of males and females who were working during the reference week. Thus, the bias could still be an important component for this type of estimate.

The adjustment for the effect of the interviewers is less important for this type of estimate because the average interviewer caseload is cut roughly in half (the average caseload for the estimate of males is about 50 percent of the caseload for all graduates). The estimated standard error of the difference goes from an unadjusted value of 0.80 percent to an adjusted value of 0.84 percent, and even this probably overestimates the increase.

Estimating the difference between domains is very similar in its statistical properties to other analytic estimates, like regressions. For many reasons Kish (1965) speculated that the impact of clustering on the standard errors would decrease for analytic estimates of differences and subdomains. The same reasoning suggests the impact of the interviewer contribution is likely to decrease for these types of estimates. The same is true for the absolute value of the bias, but not for the bias relative to the estimate. As we have seen, it is possible that the bias could have an even greater impact relative to the size of the estimate for many estimates of differences between subdomains. We will return to this point after discussing a few other examples.

Example: Certified to Teach

The certified to teach question was included in both the reconciled reinterview and the validity study. As discussed earlier in this chapter, the overall conclusion is that the estimate is probably not subject to a large measurement error bias. In addition, while there is no direct evidence of the nonresponse bias for this question, any such bias would probably be positive. This direction for the bias would be consistent with the positive bias estimated for education majors in Chapter 2.

Based on the data available, the estimates for this question would be subject to nearly the same types of adjustments as discussed above for those working for pay in the last week. As with that example, we would recommend using conservative inference procedures rather than using a poorly estimated bias adjustment.

Example: Enrollment After the Degree

One of the few questions that had a net difference rate from the reconciled reinterview that was significantly different from zero was the question that asked if the graduate had enrolled in school after receiving the degree for which they were sampled (question 12). The estimated percent of bachelor's recipients who originally said they had enrolled was 35 percent. The estimated net difference rate from the reconciled reinterview was -2.7 percent, with an estimated standard error of 1.0 percent.

Here again, adjusting this estimate based on the reinterview results would probably be inappropriate. The net difference rate shows that a higher proportion of graduates said they were enrolled when questioned in the reinterview than in the original interview. However, a substantial fraction of the difference could well be due to the time difference between the original and the reinterview rather than biases in the conditions at the time of the original interview. Using the result to adjust the estimate for bias is probably unwise.

RECOMMENDATIONS The examples in the preceding section highlight the difficulties associated with adjusting the estimates from the survey, especially for bias from nonresponse and measurement error. One of the most important concerns is the ability to estimate the biases. There are few data sources that satisfy all the requirements for use as an external source in bias evaluations. If such data sources existed and were relatively free of error, then they could be used to estimate nonresponse bias. Without these sources, the estimation of the nonresponse bias is very difficult and would be based largely on speculation.

Response bias can be estimated from reinterview studies. However, the RCG:91 reinterview was a questionable source of response bias because it was not clear that the reconciled value could be considered the correct response. A different type of reinterview could be used to address this problem by using a variety of techniques, such as different probes and more highly trained interviewers. While this type of reinterview might improve the estimation of response bias, it is generally incompatible with measuring simple response variance. For many surveys, including the RCG:91, the estimates of the simple response variance are very important in assessing the reliability of the questions for designing future interviews.

Even if the assumptions for estimating response bias are satisfied, the use of the net difference rate to adjust the estimates is a questionable practice. As discussed earlier, the adjustment affects both the estimate and its standard error. The variance of the adjusted estimate is the sum of the variance of the unadjusted estimate, the variance of the bias adjustment, and two times the covariance between the two estimates. This can be written as:

$$Var(\hat{y}_{adj}) = Var(\hat{y}) + Var(\hat{b}) + 2\ Cov(\hat{y},\hat{b}) \qquad (6.6)$$

where $\hat{b}$ is the adjustment for the bias of the estimate.

Depending on the size of the covariance term, the standard error of the adjusted estimate could be many times larger than the original estimate. As a rough rule, the sample size for the adjustment should be at least one-third of the original sample size.[9] Otherwise, the variance of the adjusted estimate is dominated by the variance of the estimated adjustment. For the RCG:91, the reinterview was conducted with only 500 of the 12,000 original respondents.

Given the problems associated with estimating the bias of the estimates from the RCG:91, we recommend analysts use the survey estimates without bias adjustments. The adjustment of the standard errors for the measurement error introduced by interviewers has fewer problems and can be recommended. The standard error estimates can be multiplied by the appropriate factors from Table 4-2. Because of the low intra-interviewer correlation, these adjustments are small or moderate for many estimates.

[9]The interview and reinterview can be considered as a two-phase or double sample to obtain better estimates of the required size of the reinterview for these types of adjustment purposes.

The other procedure recommended is the use of more conservative inference procedures, such as using 99 percent confidence intervals in place of 95 percent intervals. These conservative methods will increase the probability of estimating confidence intervals that cover the population value. Users can also take advantage of the findings from the various assessments presented in earlier chapters to determine which estimates are subject to substantial nonsampling errors. Conservative statistical procedures are recommended for those estimates most affected by nonsampling errors. For estimates not likely to be affected and for different types of exploratory analyses, these conservative procedures may not be needed.

REFERENCES AND RELATED BIBLIOGRAPHY

References and Related Bibliography

American Institutes for Research. (1992). National dropout statistics field test evaluation. (Prepared under contract to the National Center for Education Statistics, NCES 92-051). Washington, DC: U.S. Department of Education, National Center for Education Statistics.

Anderson, R., Kasper, J., Frankel, M., and Associates. (1979). *Total survey error -- Applications to improve health surveys.* San Francisco, CA: Jossey-Bass Publishers.

Bailar, B. (1968). Recent research in reinterview procedures. *Journal of the American Statistical Association,* 63(1), 41-63.

Bailar, B.A., and Biemer, P.P. (1984). Some methods for evaluating nonsampling error in household censuses and surveys. In P.S.R.S. Rao and J. Sedransk (eds.), *W.G. Cochran's impact on statistics,* 253-274. New York: John Wiley and Sons.

Biemer, P.P., and Forsman, G. (1992). On the quality of reinterview data with application to the current population survey. *Journal of the American Statistical Association,* 87(420).

Biemer, P., and Stokes, S.L. (1991). Approaches to the modeling of measurement error. In P.B. Biemer et al. (eds.), *Measurement errors in surveys,* 487-516. New York: John Wiley and Sons.

Brick, J.M., and West, J. (1992). *Reinterview program for the 1991 National Household Education Survey.* (Prepared under contract to the U.S. Department of Education.) Paper presented to the annual meeting of the American Statistical Association, Boston.

Bushery, J.M., Royce, D., and Kaspryzk, D. (1992). *The Schools and Staffing Survey: How reinterview measures data quality.* Washington, DC: U.S. Department of Commerce, Bureau of the Census.

Byce, C. (1993). *Quality of responses in the 1987 National Postsecondary Student Aid Study.* (Prepared under contract to the National Center for Education Statistics). Berkeley, CA: MPR Associates, Inc.

Cahalan, M., et al. (1993). *Occupational and educational outcomes of recent college graduates 1 year after graduation: 1991.* (Prepared under contract to the U.S. Department of Education, Office of Educational Research and Improvement, NCES 93-162.) Rockville, MD: Westat, Inc.

Cochran, W.G. (1977). *Sampling techniques.* 3rd edition. New York: John Wiley and Sons.

DeMaio, T.J. (1980). Refusals: Who, where, and why? *Public Opinion Quarterly,* 223-233.

Flemming, E. (1992). *NCES statistical standards.* U.S. Department of Education, Office of Educational Research and Improvement, NCES 92-021.

Forsman, G., and Schreiner, I. (1991). The design and analysis of reinterview - an overview. In P. Biemer et al. (eds.), *Measurement errors in surveys,* 279-302. New York: John Wiley and Sons.

Groves, R.M. (1989). *Survey errors and survey costs.* New York: John Wiley and Sons.

Groves, R.M., and Magilavy, L.J. (1980). Estimates of interviewer variance in telephone surveys. *Proceedings of the Section on Survey Research Methods.* American Statistical Association, 622-627.

Groves, R.M., and Magilavy, L.D. (1986). Measuring and explaining interviewer effects in centralized telephone surveys. *Public Opinion Quarterly*, 50(2), 251-266.

Hansen, M.H., Hurwitz, W.N., and Bershad, M.A. (1961). Measurement errors in censuses and surveys. *Bulletin of the International Statistical Institute,* 38(2), 359-374.

Hansen, M.H., Hurwitz, W.N., and Pritzker, L. (1964). The estimation and interpenetration of gross differences and the simple response variance. In C.R. Rao (ed.), *Contributions to statistics*, 111-136. Calcutta: Pergamon Press, Ltd.

Hansen, M.H., Hurwitz, W.N., Marks, E.S., and Mauldin, W.P. (1951). Response errors in surveys. *Journal of the American Statistical Association,* 46, 147-190.

Hanson, R.H., and Marks, E.S. (1958). Influence of the interviewer on the accuracy of survey results. *Journal of the American Statistical Association,* 53, 635-655.

Herzog, A.R., and Rodgers, W.L. (1988). Age and response rates to interview sample surveys. *Journal of Gerontology: Social Sciences,* 43(6), S200-205.

Hogue, C.R. (1991). *Memorandum: Evaluation study for the early estimates survey.* Washington, DC: U.S. Department of Commerce, Bureau of the Census.

Kaufman, P., and Rasinski, K.A. (1991). *Quality of the responses of eighth-grade students in NELS:88.* (Prepared under contract to the National Center for Education Statistics). Berkeley, CA: MPR Associates, Inc., and Chicago, IL: NORC.

Kish, L. (1962). Studies of interviewer variance for attitudinal variables. *Journal of the American Statistical Association,* 57, 92-115.

Kish, L. (1965). *Survey sampling.* New York: John Wiley and Sons.

Mahalanobis, P.C. (1946). Recent experiments in statistical sampling in the Indian Statistical Institute. *Journal of the Royal Statistical Society,* 109, 325-378.

Marks, E., and Mauldin, W. (1950). Response errors in Census research. *Journal of the American Statistical Association,* 45(251), 424-438.

Miller, V.P., and Groves, R.M. (1985). Matching survey responses to official records: An exploration of validity in victimization reporting. *Public Opinion Quarterly,* 49(3).

National Center for Education Statistics. (1984). *High school and beyond: A national longitudinal study for the 1980's.* Washington, DC: U.S. Department of Education, National Center for Education Statistics.

Pannekoek, J. (1988). Interviewer variance in a telephone survey. *Journal of Official Statistics*, 4, 375-384.

Peng, S. (1979). *HEGIS post-survey validation study.* Summary Report for the National Center for Education Statistics, contract number OEC-300-78-0350.

Reifman, L. (1993). *Memorandum: Statistical evaluation practices at the National Center for Education Statistics.* Washington, DC: U.S. Department of Education.

Royce, D. (1992). *1991 Schools and Staffing Survey (SASS) reinterview response variance report:* Draft. Washington, DC: U.S. Department of Commerce, Bureau of the Census.

Russ-Eft, D. (1981). *Methodological considerations for the development of error profiles of two NCES surveys.* (Prepared under contract to National Center for Education Statistics). Palo Alto, CA: American Institutes for Research.

Stokes, S.L. (1988). Estimation of interviewer effects for categorical items in a random digit dial telephone survey. *Journal of the American Statistical Association,* 83, 623-630.

Stokes, S.L., and Mulry, M.H. (1987). On the design of interpretation experiments for categorical data items. *Journal of Official Statistics,* 3, 389-401.

Temple University. *NSF study of nonrespondents to New Entrants Survey 1986-1987.* Unpublished tabulations.

U.S. Bureau of the Census. (1975). *Accuracy of data for selected housing characteristics as measured by reinterviews.* Evaluation and Research Program, Series PHC(E)-10. Washington, DC: U.S. Department of Commerce.

U.S. Bureau of the Census. (1985). *Evaluation of censuses of population and housing.* (STD-ISP-TR-5). Washington, DC: U.S. Government Printing Office.

Weaver, C.N., Holmes, S.L., and Glenn, N.D. (1975). Some characteristics of inaccessible respondents in a telephone survey. *Journal of Applied Psychology*, 60(2), 260-262.

Wolter, K. (1985). Introduction to variance estimation. Springer-Verlag.

APPENDIX A

LOCATING AND INTERVIEWING GRADUATES

APPENDIX A

LOCATING AND INTERVIEWING GRADUATES

Locating and Tracing RCG Graduates

For RCG:91 a number of procedures were used to locate graduates to be interviewed. Some of these procedures were conducted prior to survey data collection, but most were conducted during data collection. Once data collection began, 36 percent of the sample required tracing. Of the cases that required tracing, 72 percent were located. The following locating activities were conducted:

- **Survey Flyer.** Once the sample was drawn, all graduates were mailed a survey flyer enlisting the graduate's cooperation and requesting the return of an address verification form. A response from the graduate or the post office was received for 45 percent of the graduates: completed flyers were received for 25 percent; undeliverables with new addresses were received for 10 percent; and undeliverables without a new address were received for 10 percent.

- **Alumni Office Information.** The alumni offices were an important source of graduate information. They were also one of the few sources that could provide name changes. Through mail and telephone collection procedures, 93 percent of the alumni offices provided some graduate information.

- **National Change of Address (NCOA) Service.** The NCOA database is created from change of address forms submitted to the U.S. Postal Service by individuals, families, and businesses. New addresses were obtained from NCOA for about 15 percent of the graduates.

- **Referrals and Leads.** One of the best tracing sources was information from people who knew the graduate (parents, former roommates, etc.). When calling one of the telephone numbers available for a graduate, the interviewer first determined whether the graduate resided there. If not, the interviewer asked whether the respondent had any information that would help us contact the graduate. This information was very useful in tracing the graduate.

- **Telephone Tracing.** Telephone tracers searched for graduates' telephone numbers using directory assistance, referrals, and leads.

- **Credit Bureau Information.** Names of graduates that could not be located through any other procedures were sent to a professional tracing service to be matched against credit bureau information. In all, 1,462 cases were sent to the credit bureau. The credit bureau supplied addresses for 1,065 (73 percent of those sent). However, since only addresses could be supplied, tracing staff had to search for telephone numbers. A total of 389 good telephone numbers found (27 percent of those sent).

- **Telematch.** Telematch is a computerized search service that provides telephone numbers based on name, address, and ZIP code. It should be noted that Telematch did not provide new or updated address information, only phone numbers for graduates for whom we had the correct address.

Survey Data Collection Procedures

In previous RCG surveys, data collection was conducted using mail with telephone followup. The 1991 survey conducted data collection primarily by telephone, using the computer assisted telephone interviewing (CATI) system. In RCG:91, collection of questionnaires by mail was used only for graduates with unlisted numbers, those without telephones, and telephone refusals. A total of 124 surveys were completed by mail in RCG:91. Using the telephone as the primary data collection mode allowed earlier identification of graduates needing tracing and reduced the need for data retrieval.

Interviewer training was conducted during the last 3 weeks of July 1991. More than 100 interviewers were trained for the study, in groups of about 25. Each group received 16 hours of training related to the conduct of RCG:91, in addition to basic training in general interviewing techniques and the use of the CATI system. Interviewer training was conducted using the CATI system throughout. This was followed by "live" sessions that were closely monitored by training staff and telephone interviewing supervisors.

Before beginning interviewing, it was necessary to obtain telephone numbers for as many graduate addresses as possible. Telephone numbers as well as addresses had been requested from registrars, alumni offices, and graduates (through survey flyers). However, some registrars and alumni offices did not supply telephone numbers, and new addresses from NCOA and the post office did not include phone numbers. As discussed previously, a service called Telematch was used to obtain these phone numbers as quickly and efficiently as possible.

Once the address file had been updated by Telematch, all graduate information was loaded into the CATI data collection system and telephone tracing and interviewing began. Any graduate for whom no telephone number had been found went immediately into the tracing operation. As telephone data collection continued, graduates who were not located at the telephone numbers in the system also went into the tracing operation.

In order to obtain the highest possible response rate, no maximum number of calls per graduate was set. However, after seven calls, the case was reviewed by a telephone supervisor to determine the best contact approach for the case. These seven calls were staggered on different days of the week and at different times of the day over a 2-week period. The CATI system scheduled all cases automatically based on an algorithm that was customized for the RCG:91 survey.

Refusal conversion efforts were used to obtain responses from individuals who had initially refused to complete an interview. However, if the interviewer indicated that the response was "hostile" (e.g., profane or abusive), the case was reviewed by a supervisor to determine whether another attempt should be made. No more than one telephone refusal conversion attempt was made for each refusal. A 2-week hold was placed on initial refusals before a conversion attempt was made. At the end of the data collection period, a refusal conversion letter and questionnaire to be completed and returned by mail was sent to each final refusal that had a valid mailing address.

Several more procedures were followed to obtain responses from graduates who were difficult to reach by telephone, as discussed below.

- **Answering machine messages.** The first procedure involved leaving messages on graduates' answering machines, asking them to call the toll-free number. This was only done for graduates that could not be reached after repeated calls.

- **Followup letter.** The second procedure was a followup letter sent to all nonrespondents (except refusals). This letter emphasized the importance of the study and requested that the graduate call the study's toll-free number. This letter, along with the answering machine messages, helped obtain responses from graduates who were willing to participate but had schedules that made them difficult to reach.

- **Mail questionnaire.** The third procedure was to send a questionnaire to be returned by mail to graduates with unlisted numbers, those without telephones, and refusals with addresses. Graduates with unlisted telephone numbers were identified during the tracing operation through directory assistance. Those without telephone numbers were identified by a relative or friend as having no phone. Questionnaires were mailed to 1,150 graduates. Of those mailed, completed questionnaires were obtained from 11 percent.

APPENDIX B

REINTERVIEW QUESTIONNAIRE

1991 Survey of 1989-90 College Graduates Evaluation Study

[November 4, 1994]

VERIFICATION OF INFORMATION

May I speak with [STUDENT NAME]?

1. SPEAKING WITH GRADUATE (CONTINUE WITH THE INTERVIEW)
2. GRADUATE AVAILABLE (IS BEING CALLED TO THE PHONE)
3. GRADUATE NOT AVAILABLE (MAKE AN APPOINTMENT)
4. GRADUATE KNOWN BUT LIVES AT ANOTHER NUMBER
5. RECORDING-NUMBER CHANGED, DISCONNECTED OR NOT IN SERVICE
6. NEVER HEARD OF GRADUATE
7. GO TO RESULT

1. Hello, my name is {NAME} and I am calling on behalf of the United States Department of Education in regard to a study of Recent College Graduates.

EVALNTRO: Recently you participated in a study of recent college graduates for the United States Department of Education. At this time, I would like to thank you for your participation. In order to test our procedures, we are contacting a randomly selected sample of graduate participants and re-asking them a small portion of the survey.

1. CONTINUE WITH THE INTERVIEW
2. WILL NOT CONTINUE

MAJOR/GRADE POINT

IF DEGREE IS BACHELOR'S ASK Q.6: IF DEGREE IS MASTER'S OR BOTH ASK Q.9A.

6. What was your major field of study for your 1989-90 {BACHELOR'S/MASTER'S} degree? [CODE ONLY ONE: IF RESPONDENT STATES FIELD NOT **VERBATIM** ON LIST, CODE 91 OTHER]

1 ACCOUNTING
2 ANIMAL SCIENCE
3 AGRICULTURE
4 ARCHITECTURE
5 BANKING OR FINANCE
6 BIOLOGY
7 BUSINESS ADMINISTRATION
8 BUSINESS MANAGEMENT
9 BUSINESS OR MANAGEMENT
10 CHEMISTRY
11 CIVIL ENGINEERING
12 COMMUNICATIONS
13 COMPUTER SCIENCE OR INFORMATION SCIENCE
14 CRIMINAL JUSTICE OR CRIMINOLOGY
15 ECONOMICS
16 EDUCATION
17 ELEMENTARY EDUCATION
18 ELECTRICAL ENGINEERING
19 ENGLISH
20 FRENCH
21 HISTORY
22 HOME ECONOMICS
23 LIBRARY SCIENCE
24 MARKETING MANAGEMENT OR RESEARCH
25 MATHEMATICS OR ORSTATISTICS
26 MECHANICAL ENGINEERING
27 MUSIC
28 NURSING
29 PHYSICS
30 PHYSICAL EDUCATION
31 POLITICAL SCIENCE OR GOVERNMENT
32 PSYCHOLOGY
33 SOCIAL WORK
34 SOCIOLOGY
35 SPANISH
36 SPECIAL EDUCATION
37 ZOOLOGY
91 OTHER (SPECIFY)________

9A. What was your major field of study at the **undergraduate** level? [CODE ONLY ONE: IF THE RESPONDENT STATES A FIELD THAT IS NOT **VERBATIM** ON THE LIST, CODE OTHER AND SPECIFY]

1 ACCOUNTING
2 ANIMAL SCIENCE
3 AGRICULTURE
4 ARCHITECTURE
5 BANKING OR FINANCE
6 BIOLOGY
7 BUSINESS ADMINISTRATION
8 BUSINESS MANAGEMENT
9 BUSINESS OR MANAGEMENT
10 CHEMISTRY
11 CIVIL ENGINEERING
12 COMMUNICATIONS
13 COMPUTER SCIENCE OR INFORMATION SCIENCE
14 CRIMINAL JUSTICE OR CRIMINOLOGY
15 ECONOMICS
16 EDUCATION
17 ELEMENTARY EDUCATION
18 ELECTRICAL ENGINEERING
19 ENGLISH
20 FRENCH
21 HISTORY
22 HOME ECONOMICS
23 LIBRARY SCIENCE
24 MARKETING MANAGEMENT OR RESEARCH
25 MATHEMATICS OR ORSTATISTICS
26 MECHANICAL ENGINEERING
27 MUSIC
28 NURSING
29 PHYSICS
30 PHYSICAL EDUCATION
31 POLITICAL SCIENCE OR GOVERNMENT
32 PSYCHOLOGY
33 SOCIAL WORK
34 SOCIOLOGY
35 SPANISH
36 SPECIAL EDUCATION
37 ZOOLOGY
91 OTHER (SPECIFY)________

10.* On a 4-point scale, what was your grade point average for all your coursework for your undergraduate degree? [READ LIST ONLY AS PROBE: Did you receive...]

3.75-4.00 GPA (MOSTLY A'S)	1
3.25-3.74 GPA (ABOUT HALF A'S AND HALF B'S)	2
2.75-3.24 GPA (MOSTLY B'S)	3
2.25-2.74 GPA (ABOUT HALF B'S AND HALF C'S)	4
1.75-2.24 GPA (MOSTLY C'S)	5
1.25-1.74 GPA (ABOUT HALF C'S AND HALF D'S)	6
LESS THAN 1.25 (MOSTLY D'S OR BELOW)	7
HAVE NOT TAKEN COURSES FOR WHICH GRADES WERE GIVEN	8

ADDITIONAL EDUCATION

BINTRO: Throughout this questionnaire we will be referring to your {DEGREE} from {INSTITUTION} as your 1989-90 degree. Even if you have other degrees please answer only for this degree whenever we say your 1989-90 degree.

The next questions cover any additional education you may have received since obtaining your degree.

11.* During or after completing your 1989-90 degree, did you apply to any school for additional formal training?

YES 1
NO 2

12.* Have you attended school at any time since receiving the 1989-90 degree?

YES .... 1 [GO TO Q.15]
NO 2

> IF Q.11 = NO AND Q.12 = NO, ASK Q.13 AND GO TO Q.23. IF Q.11 = YES AND Q.12 = NO, DON'T ASK Q.13 AND GO TO Q.23.

13.* Which of the following best describes your reason for not applying to school? Would you say.....

You had no plans to continue your education, ...	1	[GO TO Q.23]
You wanted to work before continuing your education, ...	2	[GO TO Q.23]
You wanted to take time off before continuing your education, or	3	[GO TO Q.23]
You could not afford to continue your education?	4	[GO TO Q.23]
OTHER (SPECIFY)________________________________	91	[GO TO Q.23]

15.* Are you still enrolled?

YES .... 1 [GO TO Q.17]
NO 2

EMPLOYMENT EXPERIENCE

The next questions cover your employment experience during the week of April 22, 1991.

IF Q.12, ENROLLED, = 1, THEN ASK Q.23A; ELSE GO TO Q.23

23A.* During the week of April 22, 1991, did you have any kind of assistantships or participate in the College Work Study Program?

YES 1
NO 2

23.* Please think back to April 22 1991. Were you working for pay during this week? Please include any paid job from which you were on leave or vacation. Exclude graduate student assistantships and work study.

YES .... 1 [GO TO CINTRO]
NO 2

24.* Were you **looking** for work during the week of April 22, 1991?

YES 1
NO 2

25.* Were you **available** for work during the week of April 22, 1991?

YES 1
NO 2

26. What was the main reason you were not working during the week of April 22, 1991?

I WAS GOING TO SCHOOL (INCLUDES ASSISTANTSHIP AND WORKSTUDY) .... 1 [GO TO Q.50]
I HAD FAMILY RESPONSIBILITIES (PARENTS, CHILDREN, PREGNANCY) .... 2 [GO TO Q.50]
I COULD NOT FIND THE KIND OF JOB I WANTED .... 3 [GO TO Q.50]
I DID NOT WANT TO WORK .... 4 [GO TO Q.50]
I HAD ALREADY SECURED A NEW JOB TO BEGIN SOMETIME AFTER APRIL 22, 1991 (INCLUDING JOBS STARTING IN THE SUMMER OR FALL) .... 5 [GO TO Q.50]
I WAS LAID OFF .... 6 [GO TO Q.50]
RETIRED .... 7 [GO TO Q.50]
OTHER REASON (SPECIFY) .... 91 [GO TO Q.50]

CINTRO: Please answer the following questions for the principal job you held during the week of April 22, 1991. If you had more than one job at the same time, answer for the job from which you earned the most income, excluding assistantships and work study.

28.* What type of work were you doing? (FOR EXAMPLE: REGISTERED NURSE, ELECTRICAL ENGINEER, ACCOUNTANT, SCHOOL GUIDANCE COUNSELOR, SCHOOL TEACHER.)

Q28VERIFY [READ IF NECESSARY: WAS THE JOB RECORDED ABOVE THAT OF A SCHOOL TEACHER AT ANY GRADE LEVEL FROM PREKINDERGARTEN THROUGH GRADE 12?. EXCLUDE TUTORS, COLLEGE TEACHERS, AND DAY CARE WORKERS WITH LITTLE OR NO INSTRUCTIONAL DUTIES]

YES 1
NO 2

32.* Was this job full-time or part-time during the week of April 22, 1991?

FULL-TIME 1
PART-TIME 2

34.* Were you:

An employee of a corporatation, **private** company, business, or individual, for wages, salary, or commissions,.......... 1 [GO TO Q.38]
A **federal** government employee,.. 2 [GO TO Q.38]
A **state** government employee,.. 3 [GO TO Q.38]
A **local** government employee (city, county, etc.), or.. 4 [GO TO Q.38]
Self-employed in your **own** business, professional practice or firm?.. 5

36.* How many hours per week did you work in your business?

HOURS .... ☐☐

37.* What was your personal annual income from your business before taxes? [ENTER IN DOLLARS]

INCOME .... ☐☐☐☐☐☐.☐☐ [GO TO Q.40]

38.* How many hours per week were you usually employed at this job?

HOURS .... ☐☐

***LOGIC: CHECK AGAINST Q.32, JOBHRS MUST BE > = 30 IF FULLTIME; < 30 IF PARTTIME

39.* At what rate (before deductions) were you paid on this job?

AMOUNT .... ☐☐☐☐☐☐.☐☐ [GO TO Q.40]

PER

HOUR 1
DAY 2
WEEK 3
MONTH 4
YEAR 5

*** LOGIC: IF 0 < = JOBRATE < = 1.99, THEN JOBUNIT CANNOT = 2, 3, 4, 5.
IF 2.00 < = JOBRATE < = 9.99, THEN JOBUNIT CANNOT = 3, 4, 5.
IF 10.00 < = JOBRATE < = 39.99, THEN JOBUNIT CANNOT = 4, 5.
IF 40.00 < = JOBRATE < = 499.99, THEN JOBUNIT CANNOT = 5.
IF JOBUNIT = 1 AND JOBRATE > 100.00, THEN VERIFY. HARD RANGE = 500.00.
IF JOBUNIT = 2 AND JOBRATE > 1,000.00, THEN VERIFY. HARD RANGE = 5,000.00
IF JOBUNIT = 3 AND JOBRATE > 5,000.00, THEN VERIFY. HARD RANGE = 10,000.00
IF JOBUNIT = 4 AND JOBRATE > 20,000.00, THEN VERIFY. HARD RANGE = 30,000.00 ***

SECOND JOB

40. In addition to the principal job you have already described, were you **working for pay** at a <u>second</u> job during the week of April 22, 1991?

YES 1
NO .... 2 [GO TO Q.42]

41. Was the second job that of a school teacher at any grade level from prekindergarten through grade 12? [EXCLUDE JOB AS TEACHER'S AIDE OR DAY CARE CENTER WORKER WITH NO INSTRUCTIONAL DUTIES. ALSO EXCLUDE STUDENT TEACHING AND TUTORING]

YES 1
NO 2

RELATIONSHIP OF YOUR DEGREE TO YOUR JOB

42. Was a 4-year college degree required in order to obtain your principal job during the week of April 22, 1991?

YES 1
NO 2

43. To what extent was your work on this principal job related to your major field of study for your 1989-90 degree. Was it ...

Closely related, ... 1 [GO TO Q.45]
Somewhat related, or .... 2 [GO TO Q.45]
Not related.. 3

44. What was the <u>main</u> reason you took a job not related to your field of study?

COULD NOT FIND A JOB IN FIELD/NEEDED JOB FOR EXPENSES 1
PAY WAS BETTER 2
BETTER OPPORTUNITY FOR ADVANCEMENT 3
WANTED TO SEE IF LIKED THIS KIND OF WORK... 4
JOB WAS HELD PRIOR TO COMPLETING YOUR DEGREE 5
WANTED TO WORK IN A "MANUAL" OCCUPATION 6
BETTER OPPORTUNITY TO HELP PEOPLE OR BE USEFUL TO SOCIETY 7
OTHER (SPECIFY) 91

45. Which of the following statements <u>best</u> describes the principal job you held on April 22, 1991 with regard to career potential?

A job with definite career potential, ... 1
A job with possible career potential, or... 2
A temporary or permanent job without much career potential? 3

TEACHER CERTIFICATION AND EMPLOYMENT

DINTRO: The next questions have to do with teacher eligibility, certification, and employment. In this study I will be asking separate questions about eligibility to teach and about certification to teach.

50.* Are you eligible to teach school at any grade level from prekindergarten through grade 12? That is, have you completed all coursework, including student or practice teaching, required for a regular or standard license to teach in at least one State?

YES	1	
NO	2	[GO TO Q.53]

52.* When did you first become eligible for a certificate or license?

BEFORE JULY 1, 1989	1
JULY 1, 1989 - JUNE 30, 1990	2
AFTER JUNE 30, 1990	3

53.* Do you hold any type of regular or temporary teaching certificate or license to teach school at any grade level(s), prekindergarten through grade 12, in at least one State? [INCLUDE INITIAL, REGULAR OR STANDARD, PROVISIONAL, EMERGENCY, PROBATIONARY, OR TEMPORARY]

YES	1
NO	2

IF NO OR DON'T KNOW OR REFUSED TO Q.50 AND Q.53 SKIP TO Q.61; ELSE SKIP TO Q.58.

55. In what month and year did you first receive a certificate or license to teach?

MONTH: ☐☐ YEAR: 19☐☐

LOGIC: CERT[YY,MM] < = SYSTEM DATE

57A. Is your certification or license issued by a state?

YES	1	
NO	2	[GO TO Q.57C]

INTERVIEWER WILL ENTER 2-CHARACTER STATE, FOR Q.57B OR Q.57C. A CONFIRMATION MESSAGE WILL APPEAR WITH STATE NAME.

57B. What is the name of the state from which you received your most recent certificate or license?

_______________ STATE [GO TO Q.58]

57C. What is the name of the teacher certification agency from which you received your most recent certificate or license?

_______________ NAME OF LOCAL CERTIFICATION AGENCY

In the state of _______________

I will be reading a list of subject fields. Please tell me in which fields you have specific subject eligibility and/or certification to teach.

[FOR EACH FIELD ASK]

58. Do you have specific subject eligibility to teach? [BY ELIGIBILITY WE MEAN YOU HAVE COMPLETED ALL COURSEWORK, INCLUDING STUDENT OR PRACTICE TEACHING, REQUIRED FOR A REGULAR OR STANDARD LICENSE TO TEACH IN AT LEAST ONE STATE. [SKIP IF Q.50 IS NO]

59. Do you have specific subject certification to teach [BY CERTIFIED WE MEAN YOU HOLD SOME TYE OF REGULAR OR TEMPORARY TEACHING CERTIFICATE OR LICENSE TO TEACH SCHOOL AT ANY GRADE LEVEL, PREKINDERGARTEN THROUGH GRADE 12, IN AT LEAST ONE STATE.] [SKIP IF Q.53 IS NO]

		Q.58 Column A. Fields eligible to teach		**Q.59 Column B.** Fields certified to teach	
		Yes	No	Yes	No
1.	Any Elementary fields, general or specialized	1	2	1	2
2.	Art/fine art/performing arts	1	2	1	2
3.	Basic skills and remedial education	1	2	1	2
4.	Bilingual education	1	2	1	2
5.	Biological or life sciences	1	2	1	2
6.	Business (not part of voc. ed. curriculum)	1	2	1	2
7.	Computer science	1	2	1	2
8.	English language arts	1	2	1	2
9.	English-as-a-second language	1	2	1	2
10.	Foreign languages	1	2	1	2
11.	Gifted/talented	1	2	1	2
12.	Health	1	2	1	2
13.	Home economics	1	2	1	2
14.	Industrial Arts, Trade, and Industry	1	2	1	2
15.	Mathematics	1	2	1	2
16.	Music	1	2	1	2
	Any Physical sciences, general or specialized: [IF YES ASK]				
17.	General Sciences (no specialized area)	1	2	1	2
18.	Chemistry	1	2	1	2
19.	Geology/earth science	1	2	1	2
20.	Physics	1	2	1	2
21.	Other physical sciences	1	2	1	2
22.	Physical education	1	2	1	2
23.	Pre-elementary education	1	2	1	2
24.	Reading	1	2	1	2
25.	Religion/philosophy	1	2	1	2
26.	Social science/social studies	1	2	1	2
	Any Special education fields [IF YES ASK;]				
27.	Mentally retarded	1	2	1	2
28.	Hearing impaired, deaf	1	2	1	2
29.	Seriously emotionally disturbed	1	2	1	2
30.	Speech impaired	1	2	1	2
31.	Specific learning disability	1	2	1	2
32.	General certificate (no specific condition)	1	2	1	2
33.	Other special education	1	2	1	2
34.	Vocational Education, other than Business, Home Economics, or Industrial Arts	1	2	1	2
35.	Other fields [INCLUDES GENERAL SECONDARY CERTIFICATE]	1	2	1	2

SKIP Q.61 IF Q28VERIFY = 1, SET EVERTEAC TO 1

61. Have you ever taught any grade from prekindergarten through grade 12?

YES .... 1 [GO TO Q.62]
NO 2

IF NO, REFUSED, OR DON'T KNOW TO Q.61 SET Q.62 = 2 AND SKIP TO Q.64

62. Prior to completing the requirements for your 1989-90 degree, were you at any time employed as a school teacher at any grade level, from prekindergarten through grade 12? Please exclude student or practice teaching and work as a teacher's aide.

YES 1
NO 2

APPLIED FOR A TEACHING POSITION

Now I would like to ask you about applying for teaching positions.

64. Have you applied for a job as a school teacher at any grade level from prekindergarten through grade 12 since or immediately prior to receiving your 1989-90 degree?

YES	1
NO	2

> IF NO, REFUSED, OR DON'T KNOW TO Q.50, Q.53, AND Q.61, SET Q.65 TO 1 AND SKIP TO Q.94; ELSE IF Q.64 = 1, GO TO Q.66

65. What was the main reason you decided not to apply for a teaching job?

NEVER INTERESTED IN TEACHING	1
MORE EDUCATION BEFORE TEACHING (NOT READY)	2
HAD ALL COURSEWORK NEEDED BUT NOT READY TO APPLY	3
DID NOT BOTHER TO APPLY BECAUSE JOBS ARE HARD TO GET	4
STUDENT TEACHING EXPERIENCE DISCOURAGED ME	5
MORE MONEY IN OTHER JOB OFFER	6
MORE PRESTIGE IN OTHER JOB OFFER	7
WANTED OTHER OCCUPATION (SPECIFY OCCUPATION)________	8
LOW PAY	9
TEACHING CONDITIONS	10
ALREADY HAD A TEACHING JOB	11
OTHER (SPECIFY) ________	91

> SKIP Q.66 IF Q28VERIFY = 1; SET EVEDEGR TO 1

IF (Q.50 = 1 OR Q53 = 1) AND (Q.61 = 2) THEN DISPLAY Q.66 TEXT BEFORE Q.66.

66. I've recorded that you've never taught any grade from prekindergarten through grade 12. Before continuing, I'd like to verify your teaching status since receiving your 1989-90 degree.

Have you taught at any grade level, from prekindergarten through grade 12, since receiving your 1989-90 degree?

YES 1
NO .... 2 [GO TO Q.94.]

TEACHER EMPLOYMENT

The next questions have to do with your employment as a teacher.

67.* In what month and year did you first start teaching?

MONTH:................☐☐ YEAR: .... 19☐☐

LOGIC: TEACH [YY,MM] < = SYSTEM DATE

SKIP Q.68 IF Q28VERIFY = 1; SET MAINTEAC TO 1

68.* During the week of April 22, 1991 was your principal job that of a school teacher at any grade level from prekindergarten through grade 12?
[PRINCIPAL JOB MEANS THE JOB FROM WHICH YOU EARN MOST OF YOUR INCOME]

YES 1
NO .... 2 [GO TO Q.94.]

*** LOGIC: IF MAINTEAC = YES, Q28VERIFY (OCCUVERF) MUST = 1 ***

71. Please tell me all the fields in which you were teaching during the week of April 22, 1991. [CODE ALL THAT APPLY INTO GENERAL CATEGORIES LISTED BELOW.] [FOR ELEMENTARY TEACHERS,

5.71 CODE "ANY ELEMENTARY FIELDS." CODE SEPARATE FIELDS ONLY IF TEACH SEPARATE CLASSES]

NONE MUST BE ENTERED ALONE; ASK Q.72 ONLY IF INDICATED TAUGHT IN MORE THAN ONE FIELD

72. During the week of April 22, 1991, what was the field in which you taught most of the time?

		Q.71 Fields teaching	Q.72 **Code only one** Field taught most frequently
0.	NONE	00	
1.	ANY ELEMENTARY FIELDS, GENERAL OR SPECIALIZED	01	01
2.	ART/FINE ART/PERFORMING ARTS	02	02
3.	BASIC SKILLS AND REMEDIAL EDUCATION	03	03
4.	BILINGUAL EDUCATION	04	04
5.	BIOLOGICAL OR LIFE SCIENCES	05	05
6.	BUSINESS (NOT PART OF VOC. ED. CURRICULUM)	06	06
7.	COMPUTER SCIENCE	07	07
8.	ENGLISH LANGUAGE ARTS	08	08
9.	ENGLISH-AS-A-SECOND LANGUAGE	09	09
10.	FOREIGN LANGUAGES	10	10
11.	GIFTED/TALENTED	11	11
12.	HEALTH	12	12
13.	HOME ECONOMICS	13	13
14.	INDUSTRIAL ARTS/TRADE	14	14
15.	MATHEMATICS	15	15
16.	MUSIC	16	16
	ANY PHYSICAL SCIENCES, GENERAL OR SPECIALIZED:		
17.	GENERAL SCIENCES (NO SPECIALIZED AREA)	17	17
18.	CHEMISTRY	18	18
19.	GEOLOGY/EARTH SCIENCE	19	19
20.	PHYSICS	20	20
21.	OTHER PHYSICAL SCIENCES	21	21
22.	PHYSICAL EDUCATION	22	22
23.	PRE-ELEMENTARY EDUCATION	23	23
24.	READING	24	24
25.	RELIGION/PHILOSOPHY	25	25
26.	SOCIAL SCIENCE/SOCIAL STUDIES	26	26
	ANY SPECIAL EDUCATION FIELDS		
27.	MENTALLY RETARDED	27	27
28.	HEARING IMPAIRED, DEAF	28	28
29.	SERIOUSLY EMOTIONALLY DISTURBED	29	29
30.	SPEECH IMPAIRED	30	30
31.	SPECIFIC LEARNING DISABILITY	31	31
32.	GENERAL CERTIFICATE (NO SPECIFIC CONDITION)	32	32
33.	OTHER SPECIAL EDUCATION	33	33
34.	VOCATIONAL EDUCATION - OTHER	34	34
35.	OTHER FIELDS	35	35

TEACHING ASSIGNMENT

The next questions are about your teaching assignment.

85. Was your teaching assignment full-time or part-time during the week of April 22, 1991?

FULL-TIME 1
PART-TIME 2

87. Were you working under a teaching contract or did you have some other arrangement, such as substitute teaching?

TEACHING CONTRACT 1
SUBSTITUTE TEACHING .... 2 [GO TO Q.94]
INTERNSHIP .... 3 [GO TO Q.94]
OTHER (SPECIFY) .... 91 [GO TO Q.94]

87a. How many months per year was your principal teaching contract?

NUMBER OF MONTHS PER YEAR: ....☐☐

87b. How many months per year were you paid?

NUMBER OF MONTHS PER YEAR: ....☐☐

87c. What was your annual income from the principal teaching contract under which you were working on April 22, 1991?

[ENTER IN DOLLARS]
AMOUNT ....☐☐☐☐☐☐.☐☐

87d. Do you expect any other earned income from summer employment outside of your principal teaching job in 1991?

YES 1
NO .... 2 [GO TO Q.94]

87e. What is the total amount you expect to earn from summer employment?

[ENTER IN DOLLARS]
AMOUNT ....☐☐☐☐☐☐.☐☐

IF TEACXINC = 0, THEN RESET TEACXTRA TO 2

BACKGROUND INFORMATION

94. Are you of Hispanic or Spanish origin?

YES	1
NO	2

95. What race do you consider yourself?

WHITE [CAUCASIAN]	1
BLACK [AFRICAN AMERICAN]	2
NATIVE AMERICAN OR ALASKA NATIVE [AMERICAN INDIAN]	3
OTHER (SPECIFY) ____________________	91

FINANCIAL SUPPORT TO ATTEND SCHOOL

EINTRO:

The next questions concern how you financed your education. Please answer only for the {INSERT DEGREE} degree you received from {INSERT INSTITUTION} in 1989-90, and not any other education you may have received before or after receiving the 1989-90 degree.

I will be reading a list of possible sources of financial support. I would first like to identify which ones you used. We are interested in your total expenses including tuition, fees, room, board, supplies, transportation, and miscellaneous expenses.

106. In financing your 1989-90 degree, did you use....

	SOURCE OF PAYMENT	YES	NO
a.	Your own earnings and personal savings excluding work study earnings?	1	2
b.	Your earnings from work study?	1	2
c.	Support from spouse?	1	2
d.	Support from parents?	1	2
	[IF YES] Was this in the form of:		
	1. Support to be paid back (loans)	1	2
	2. Support NOT to be paid back	1	2
e.	Support from relatives or friends?	1	2
	[IF YES] Was this in the form of:		
	1. Support to be paid back (loans)	1	2
	2. Support NOT to be paid back	1	2
f.	Employers support?	1	2
	[IF YES] Was this in the form of: OVERLAY		
	1. Support to be paid back (loans)	1	2
	2. Support NOT to be paid back	1	2
g.	Loans from any source other than from parents, relatives, friends, or employers)?	1	2
h.	Grants or scholarships from Federal, State, or local government or your college or university?	1	2
i.	Grants or scholarships from any other source such as private companies or civic organizations?	1	2
j.	Fellowships from any source?	1	2
k.	Assistantships from any source?	1	2
l.	Any other sources that I have not mentioned? [SPECIFY]________________	1	2

AT LEAST 1 RESPONSE AT Q.106 MUST BE 1 OR MISSING

IF YES TO Q.106h or Q.106i, ASK FOR EACH TYPE

110. We are interested in the types of grants or scholarships you have ever received for your 1989-90 degree.

At any time while working on your 1989-90 degree did you ever have .....

Did you have this form of aid between July 1, 1989 and June 30, 1990?

		EVER		BETWEEN JULY 1, 1989 AND JUNE 30, 1990	
		YES	NO	YES	NO
a.	Federal Pell or BEOGS grants?	1	2	1	2
b.	Other Federal grants or scholarships?	1	2	1	2
c.	State grants or scholarships?	1	2	1	2
d.	Institutional grants or scholarships?	1	2	1	2
e.	Other grants or scholarships?	1	2	1	2

IF YES TO Q.106g, ASK EACH TYPE; ELSE SKIP TO BOX BEFORE Q.115.

111. Now I'd like to ask about loans other than from parents, relatives, friends, or employers. At any time while working on your 1989-90 degree, did you have any of the following types of loans? Did you ever have...

Did you have this form of aid between July 1, 1989 and June 30, 1990 year?

		EVER		BETWEEN JULY 1, 1989 AND JUNE 30, 1990	
		YES	NO	YES	NO
a.	Federal Guaranteed Student Loan (GSL) Program now called the Stafford Loan?	1	2	1	2
b.	The Supplemental Loans for Students (SLS)?	1	2	1	2
c.	Other Federal loans (Perkins, Income Contingent)?	1	2	1	2
d.	State loans?	1	2	1	2
e.	Institutional loans?	1	2	1	2
f.	Other loans excluding loans from parents, friends relatives, or employers?	1	2	1	2

AT LEAST 1 "EVER" AT Q.111 MUST EQUAL "1" IF ASKED. IF NOT, MESSAGE INTERVIEWER. IF CONFIRMS NO "1" SECOND TIME THROUGH, GO TO END.

RECONCILIATION QUESTIONS

THE FOLLOWING QUESTIONS WILL BE ASKED ONLY IF THE ANSWERS GIVEN IN THE TWO INTERVIEWS ARE DIFFERENT AND NEITHER ANSWER IS REFUSED OR DON'T KNOW.

*10R. During our original interview with you, we recorded that you had not taken courses for which grades were given/your undergraduate GPA was ___

Now I have recorded that you did not take courses for which grades were given/your undergraduate GPA was GPA2.

1. Was the original answer correct, or [GO TO Q.11R]
2. Is the new answer correct, [GO TO Q.11R]
3. Or, is neither answer correct?

OPT10 What was your undergraduate GPA?

1. 3.75 - 4.00 (MOSTLY A'S)
2. 3.25 - 3.74 (HALF A'S & HALF B'S)
3. 2.75 - 3.24 (MOSTLY B'S)
4. 2.25 - 2.74 (HALF B'S & HALF C'S)
5. 1.75 - 2.24 (MOSTLY C'S)
6. 1.25 - 1.74 (HALF C'S & HALF D'S)
7. LESS THAN 1.25 (MOSTLY D'S & BELOW)
8. DID NOT TAKE COURSE FOR GRADE

11R. During our original interview with you, we recorded that you had/had not applied to any school for additional training after completing your 1989-90 degree.

Now I have recorded that you have/have not applied for additional training after
completing your 1989-90 degree.

1. Was the original answer correct, or
2. Is the new answer correct,

** 4. Or, has the situation changed since we last spoke with you?

12R. During our original interview with you, we recorded that you had/had not attended school at any time since receiving your 1989-90 degree.

Now I have recorded that you have/have not attended school at any time since receiving your 1989-90 degree.

1. Was the original answer correct, or
2. Is the new answer correct,

** 4. Or, has the situation changed since we last spoke with you?

13R. During our original interview with you, we recorded that your <u>best</u> reason for not applying to school was that you had no plans to continue/want to work before continuing/could not afford to continue NOTATEOS your education

Now I have recorded that <u>best</u> reason for not applying to school is that you had no plans to continue/want to work before continuing/could not afford to continue NOTATOS2 your education

1. Was the original answer correct, or [GO TO Q23R]
2. Is the new answer correct, [GO TO Q23R]
3. Is neither answer correct,
4. Or, has the situation changed since we last spoke with you? [GO TO Q23R]

OPT13. What was your best reason for not applying to school?

1. You hadno plans to continue your education,
2. You wanted to work before continuing your education,
3. You wanted to take time offf before continuing your education, or
4. You could not afford to continue your education?
91. OTHER

15R. During our original interview with you, we recorded that you were/were not still enrolled in school since receiving your 1989-90 degree.

Now I have recorded that you are/are not still enrolled in school.

1. Was the original answer correct, or
2. Is the new answer correct,
** 4. Or, has the situation changed since we last spoke with you?

23AR. During our original interview with you, we recorded that you did have a kind of/did not have any kind of assistantship or/nor participated in the College Work Study Program during the week of April 22, 1991.

Now I have recorded that you did have an assistantship or participated/did not have any assistantship nor participated in the College Work Study Program during the week of April 22, 1991.

1. Was the original answer correct, or
2. Is the new answer correct?

23R. During our original interview with you, we recorded that you were/were not working for pay during the week of April 22, 1991.

Now I have recorded that you did/did not work for pay during the week of April 22, 1991.

1. Was the original answer correct, or
2. Is the new answer correct?

24R. During our original interview with you, we recorded that you were/were not looking for work during the week of April 22, 1991.

Now I have recorded that you were/were not looking for work during the week of April 22, 1991.

1. Was the original answer correct, or
2. Is the new answer correct?

25R. During our original interview with you, we recorded that you were/were not available for work during the week of April 22, 1991.

Now I have recorded that you were/were not available for work during the week of April 22, 1991.

1. Was the original answer correct, or
2. Is the new answer correct?

INTERVIEWER WILL BE ASKED:

DURING THE FIRST INTERVIEW THE RESPONDENT SAID THAT HE/SHE WAS EMPLOYED AS A/AN (OCCUPATN).

DURING THIS INTERVIEW HE/SHE WAS A/AN (OCCUPAT2).

ARE THESE TWO OCCUPATIONS THE SAME?

1. YES
2. NO [GO TO Q.28R]

28R. During our original interview with you, we recorded that you worked as a/an OCCUPATN during the week of April 22, 1991.

Now I have recorded that you worked as a/an OCCUPAT2 during the week of April 22, 1991.

1. Was the original answer correct, or [GO TO Q.32R]
2. Is the new answer correct, [GO TO Q.32R]
3. Or, is neither answer correct?

RECO28. What is the correct answer? [What type of work were you doing during the week of April 22, 1991]

CORRECT ANSWER:____

32R. During our original interview with you, we recorded that you were employed full-time/part-time during the week of April 22, 1991.

Now I have recorded that you were employed full-time/part-time during the week of April 22, 1991.

1. Was the original answer correct, or
2. Is the new answer correct?

34R. During our original interview with you, we recorded that you were an employee of a corporation, private company, business or individual/an employee of the federal government/an employee of a state government/self-employed during the week of April 22, 1991.

Now I have recorded that you were an employee of a corporation, private company, business or individual/an employee of the federal government/an employee of a state government/self-employed during the week of April 22, 1991.

1. Was the original answer correct, or [GO TO Q.36]
2. Is the new answer correct, [GO TO Q.36]
3. Or, is neither answer correct?

OPT34. Were you:

1. An employee of a corporation, private company, business or individual,
2. A federal government employee,
3. A state government employee,
4. A local government employee, or
5. Self employed in your own business, professional practice or firm?

IF Q34 = 5, THEN GO TO Q.36R ELSE GO TO Q.38R

36R. During our original interview with you, we recorded that during the week of April 22, 1991 you worked ___ hours per week at your business.

Now I have recorded that you worked ___ hours per week at your business.

1. Was the original answer correct, or
2. Is the new answer correct,
3. Or, is neither answer correct?

37R. During our original interview with you, we recorded that your personal annual income from your business before taxes was $_________ as of April 22, 1991.

Now I have recorded that your personal annual income from your business was $_________ on April 22, 1991.

1. Was the original answer correct, or [GO TO Q.38R]
2. Is the new answer correct, [GO TO Q.38R]
3. Or, is neither answer correct?

REININCM. What was your annual income from your business?

ANSWER: $ ___________,

38R. During our original interview with you, we recorded that during the week of April 22, 1991 you worked ___ hours per week.

Now I have recorded that you worked ___ hours per week.

1. Was the original answer correct, or [GO TO Q.39R]
2. Is the new answer correct, [GO TO Q.39R]
3. Or, is neither answer correct?

OPT38. How many hours per week did you work during the week of April 22, 1991?

39R. During our original interview with you, we recorded that you were paid $______ dollars per hour/day/week/month/year for the job you held during the week of April 22, 1991.

Now I have recorded that you were paid $______ dollars per hour/day/week/month/year for the job you held during that week.

1. Was the original answer correct, or [GO TO Q.40R]
2. Is the new answer correct, [GO TO Q.40R]
3. Or, is neither answer correct?

REININCM. What was your {annual/monthly/weekly/daily/hourly} income?

ANSWER: $ ___________,

PER: ________

1. HOUR
2. DAY
3. WEEK
4. MONTH
5. YEAR

40R. During our original interview with you, we recorded that you were/were not working at a second job for pay during the week of April 22, 1991.

Now I have recorded that you were/were not working at a second job for pay during the week of April 22, 1991.

1. Was the original answer correct, or
2. Is the new answer correct?

41R. During our original interview with you, we recorded that your second job was/was not that of a school teacher at any grade level from prekindergarten through grade 12.

Now I have recorded that your second job was/was not that of a school teacher.

1. Was the original answer correct, or
2. Is the new answer correct?

50R. During our original interview with you, we recorded that you were/were not eligible to teach in at least one State.

Now I have recorded that you are/are not eligible to teach in at least one State.

1. Was the original answer correct, or
2. Is the new answer correct,

** 4. Or, has the situation changed since we last spoke with you?

52R. During our original interview with you, we recorded that you first became eligible for a certificate or license before July 1, 1989/between July 1, 1989 and June 30, 1990/after June 30, 1990.

Now, I have recorded that you first became eligible for a certificate or license before July 1, 1989/between July 1, 1989 and June 30, 1990/after June 30, 1990.

1. Was the original answer correct, or [GO TO Q.53R]
2. Is the new answer correct, [GO TO Q.53R]
3. Or, is neither answer correct?

OPT52 When did you first become eligible for a certificate or license?

1. BEFORE JULY 1, 1989
2. JULY 1, 1989 - JUNE 30, 1990
3. AFTER JUNE 30, 1990

53R. During our original interview with you, we recorded that you did not have any/did have some type of regular or temporary teaching certificate or license to teach school at any grade level, prekindergarten through grade 12, in at least one State.

Now, I have recorded that you do not have any/ do have some type of regular or temporary teaching certificate or license to teach school at any grade level in at least one State.

1. Was the original answer correct, or
2. Is the new answer correct,

** 4. Or, has the situation changed since we last spoke with you?

67R. During our original interview with you, we recorded that you first started teaching in __DATE__.

Now I have recorded that you first started teaching in __DATE__.

1. Was the original answer correct, or [GO TO Q.68R]
2. Is the new answer correct, [GO TO Q.68R]
3. Or, is neither answer correct?

OPT67. In what month and year did you first start teaching?

1. JANUARY
2. FEBRUARY
3. MARCH
4. APRIL
5. MAY
6. JUNE
7. JULY
8. AUGUST
9. SEPTEMBER
10. OCTOBER
11. NOVEMBER
12. DECEMBER

68R. During our original interview with you, we recorded that your principal job during the week of April 22, 1991 was/was not that of a school teacher at any grade level from prekindergarten through grade 12.

Now I have recorded that your principal job during the week of April 22, 1991 was/was not that of a school teacher at any grade level from prekindergarten through grade 12.

1. Was the original answer correct, or
2. Is the new answer correct?

[The following question will be asked after each reconciliation question.]

OPINION What do you think might be the reason for the difference between what we recorded in the first and second interview? Was it because...

1. It was difficult to recall an exact answer to the question, or
2. The question was unclear or the response category used in the question did not fit your situation, or
3. The wrong response was recorded by our interviewer, or
4. Your perception has changed since the interview was first conducted?
91. OTHER

[READ]
TIMEBURD:
If you have any comments regarding the time burden of this survey or any other aspect of this data collection, including suggestions for reducing the time burden, you may write to the U.S. Department of Education.

[IF RESPONDENT INDICATES WOULD LIKE TO WRITE GIVE ADDRESS AS FOLLOWS]

U.S. Department of Education
Information, Management, and Compliance Division
Washington, D.C. 20202-4651

THANKYO1:

AT THIS TIME I'D LIKE TO THANK YOU VERY MUCH FOR YOUR PARTICIPATION IN THIS STUDY.

*These questions are reconciled at the end of the survey. #10 is the last question to be reconciled because of it's sensitivity.

**OPTION4 will not be allowed for Questions 11, 12, 50 and 53 only if the first answer was NO and the second answer is YES. For questin 15, Option 4 will be allowed if the first answer was YES and the second answe is NO.

APPENDIX C

SELF-REPORTED REASONS FOR DISCREPANCIES IN REINTERVIEW

APPENDIX C
SELF-REPORTED REASONS FOR DISCREPANCIES IN REINTERVIEW

Reconciliation of Response Discrepancies

The purposes of the reconciliation process were (1) to obtain the most accurate responses to selected questionnaire items for use in the estimates of bias discussed above; (2) to obtain the graduates' explanations of the most likely reason for the discrepancy; and (3) to use the graduates' explanations to identify possible problems with specific questionnaire items. This appendix focuses on graduates' explanations for the discrepancies.

Graduate Identification of the Correct Answer

Once all the reinterview questionnaire items had been asked, the CATI system compared responses from the original survey to the reinterview responses for each of the questionnaire items being reconciled. When there was a discrepancy between the response on the original and the response on the reinterview, the graduate was informed of the discrepancy and asked to identify the correct answer. The graduate was asked the following question:

During our original interview with you, we recorded that...
Now I have recorded that...

1. Was the original answer correct, or
2. Is the new answer correct, or
(3. Is neither answer correct,)
(4. Or, has the situation changed since we last spoke with you?)

Table C-1 below shows the distribution of responses to this resolution of discrepancies.

Table C-1. Resolution of response discrepancies

Resolution of discrepancies	Total		Excluding income items	
	Number	Percent	Number	Percent
Total discrepancies	899	100%	671	100%
Original answer correct	390	43	263	39
New answer correct	424	47	335	50
Neither answer correct[1]	25	3	17	3
Situation has changed[2]	40	4	40	6
Don't know	20	2	16	2

[1]This resolution was only applicable for questionnaire items with more than two response categories. This includes Q10, Q13, Q28, Q34, Q36, Q37, Q38, Q39, Q52, and Q67. Among the cases where "neither answer correct" was applicable, it was chosen 25 of 656 times (4 percent).

[2]This resolution was only applicable for questionnaire items where it was possible for the situation to change. This includes Q11, Q12, Q13, Q15, Q50, and Q53. Among the cases where "situation changed" was applicable, it was chosen 40 of 200 times (20 percent).

NOTE: Percentages may not add to 100 due to rounding.

Original Answer Correct and New Answer Correct Categories. Overall, graduates said that the original answer was correct for 43 percent of the discrepancies, and that the new (reinterview) answer was correct for 47 percent of the discrepancies. This distribution changes when the income items (Q37 and Q39) are excluded. For income (as for all questionnaire items), the original response was matched exactly to the reinterview response, and any difference between the two responses required a reconciliation with the graduate. Therefore, small differences due to rounding and differences in the reporting unit (year, month, week, day, or hour) were included as discrepancies. If the graduate indicated that the responses were actually the same, the interviewer was instructed to choose the category "original answer correct." This was done so that difference rates calculated using reconciled responses do not include these cases where the two responses are actually the same. Therefore, the "original answer correct" category is slightly inflated for the income items, and a more accurate distribution may be obtained when the income items are excluded. When income items are excluded, graduates said that the original answer was correct for 39 percent of the discrepancies, and that the new (reinterview) answer was correct for 50 percent of the discrepancies.

If the reinterview is an independent replication of the original interview, then the number of original and reinterview errors should be roughly equal. In this case the differences between the two categories (original answer correct and new answer correct) is small, indicating that the reinterview was relatively successful in producing an independent replication of the original survey.

Neither Answer Correct Category. This category was only included when more than two responses were possible for a questionnaire item. About half the reconciliation questions included this answer category. Among all discrepancies, this category was chosen 3 percent of the time. Among the questions where it was applicable, it was chosen 4 percent of the time.

Situation Has Changed Category. This category was only included for the six questionnaire items where a situation change was possible. Among all discrepancies, this category was chosen 4 percent of the time. Among the questions where it was applicable, it was chosen 20 percent of the time. Therefore, this is a significant category for some questions.

Distribution by Questionnaire Item. Table C-2 contains the distribution of resolution categories by questionnaire item. Only items with at least 20 discrepancies are included in the table, except question 53, which had 10 discrepancies and is included because it is a key question and is discussed in Chapter 5.

Table C-2. Resolution of response discrepancies by questionnaire item

Questionnaire Item**	Number of cases	Percent in each resolution category				
		Original answer correct	Reinterview answer correct	Neither answer correct	Situation changed	Don't know
Q10 GPA	121	43%	54%	2%	*	2%
Q11 Applied to school	47	30	57	*	13%	0
Q12 Attended school	47	23	47	*	30	0
Q13 Reason did not apply	71	37	42	0	18	3
Q23 Working for pay	21	48	48	*	*	5
Q32 Job full or part time	27	26	74	*	*	0
Q34 Type of employer	27	30	63	0	*	7
Q38 Hours per week employed	146	41	47	9	*	3
Q39 Salary amount (unit the same) . . .	133	51	41	6	*	2
Q39 Salary unit	93	63	34	0	*	2
Q40 Working second job	24	50	50	*	*	0
Q53 Certified to teach	10	0	60	*	40	0
Q67 Month started teaching	26	50	46	4	*	0

*Not applicable. This was not a possible response category for the questionnaire item.

**Only reconciliation questionnaire items with at least 20 discrepancies were included in this table except question 53 (certified to teach), which is included because it is a key question and is discussed in chapter 5.

NOTE: Percentages may not add to 100 due to rounding.

Reasons for Discrepancies

Within the RCG:91 reinterview, once the discrepancy had been resolved for a questionnaire item, the interviewer asked the graduate to identify the most likely reason the discrepancy had occurred. However, if the response to the resolution question had been that the situation had changed, the graduate was not asked the reason for the difference. Instead, the CATI program automatically entered a code 5 as the reason. For all other cases, the graduate was asked the following question:

> What do you think might be the reason for the difference between what we recorded in the first and second interview? Was it because:
>
> 1. It was difficult to recall an exact answer to the question, or
> 2. The question was unclear or the response category used in the question did not fit your situation, or
> 3. The wrong response was recorded by our interviewer, or
> 4. Your perception has changed since the interview was first conducted?
> 5. SITUATION HAS CHANGED
> 91. OTHER

Table C-3 shows the distribution of reasons for the response discrepancies.

Table C-3. Reasons for response discrepancies

Reason for discrepancy	Total		Excluding income items	
	Number	Percent	Number	Percent
Total discrepancies	899	100%	671	100%
It was difficult to recall an exact answer to the question	320	36	216	32
The question was unclear or the response category did not fit R's situation	183	20	154	23
The wrong response was recorded by the interviewer	98	11	90	13
R's perception changed since the interview was first conducted	114	13	88	13
The situation changed*	40	4	40	6
Other	142	16	81	12
Don't know	2	-	2	-

*This reason was only applicable for questionnaire items where it was possible for the situation to change. This includes Q11, Q12, Q13, Q15, Q50, and Q53. Among the cases where "situation changed" was applicable, it was chosen 40 of 200 times (20 percent).

- Less than 0.5 percent.

NOTE: Percentages may not add to 100 due to rounding.

Recall problems were cited by graduates as the most common reason for response discrepancies (36 percent of the time). This was followed by the question was unclear or the response category did not fit the graduate's situation (20 percent), other reasons (16 percent), the graduate's perception changed (13 percent), and the wrong response was recorded by the interviewer (11 percent).

Respondent error accounted for about one-third of the reasons included in the "other" category. Another one-third of the "other" category was that both answers were correct or both were the same. This occurred most frequently in the questionnaire items of (1) hours per week, when the hours were rounded differently; (2) occupation, when the same occupation was reported slightly differently; (3) salary, when the unit (year, month, week, day, hour) was reported differently; and (4) reason for not applying to school after the degree, when the graduate said that both reasons for not applying were correct or the two reasons were basically the same.

Table C-4 contains the distribution of reasons by questionnaire item. Only items with at least 20 discrepancies are included in the table. When examining this table, it is important to look at the percentage of cases with discrepancies as well as the percentage in the reason category. For example, both Q23 (whether working for pay) and Q39 (salary amount) have 57 percent of their discrepancies caused by recall problems. However, Q23 only has discrepancies for 4 percent of the cases, while Q39 has discrepancies for 34 percent of the cases. This means that about 2 percent of the reinterview sample had difficulty recalling whether they worked for pay (calculated as .57 x 4), and about 19 percent of the reinterview sample had difficulty recalling the exact salary amount (calculated as .57 x 34).

Question Q10. For grade point average (GPA), about 24 percent of the cases had discrepancies. Among those with discrepancies, the most common reasons were recall problems (41 percent) and interviewer error (37 percent). Recall can be affected by both the amount of time elapsed since the event and whether specific information is being requested. Most graduates had received their undergraduate degree at least 1 year before the survey, and grade point average is very specific information.

This GPA question had the highest rate of interviewer error reported. When administering this question, the interviewer asked for the specific grade point average, and then chose the correct answer category. Interviewers may have made errors in choosing the answer category. It is also possible that some discrepancies reported as interviewer errors were actually graduate errors. Graduates who did not remember their previous responses may have assumed that the interviewer made the error.

Table C-4. Reason for response discrepancies by questionnaire item.

Questionnaire item[1]	Number of discrepancies	Gross percent of cases with discrepancies[2]	Percent in each reason category						
			Recall problem	Unclear question	Interviewer error	Perception changed	Situation changed	Other	Don't know
Q10 GPA	121	24.1%	41%	8%	37%	6%	*	7%	0%
Q11 Applied to school	47	9.2	15	40	11	15	13%	6	0
Q12 Attended school	47	9.2	4	23	15	17	30	11	0
Q13 Reason did not apply	71	26.4	13	23	3	24	18	20	0
Q23 Working for pay	21	4.1	57	19	5	0	*	19	0
Q32 Job full or part time	27	6.4	19	30	7	26	*	19	0
Q34 Type of employer	27	6.5	30	41	7	22	*	0	0
Q38 Hours per week employed	146	35.7	47	16	5	15	*	16	0
Q39 Salary exact amount (unit the same)	133	34.1	57	11	5	12	*	15	0
Q39 Salary unit	93	23.8	30	16	1	10	*	43	0
Q40 Working second job	24	5.7	58	17	13	8	*	4	0
Q53 Certified to teach	10	2.0	0	20	10	20	40	10	0
Q67 Month started teaching	26	23.9	54	27	8	4	*	4	4

*Not applicable. This was not a possible response category for the questionnaire.

[1]Only reconciliation questionnaire items with at least 20 discrepancies were included in this table except question 53 (certified to teach), which is included because it is a key question and is discussed in chapter 5.

[2]The percentage of cases with discrepancies is based on the number of discrepancies for a questionnaire item divided by the number of cases for which that item was applicable and answered (i.e., not don't know or refused).

NOTE: Percentages may not add to 100 due to rounding.

Questions Q11, Q12, Q13. These three questions on school attendance after receiving the 1989-90 degree were asked as follows:

11. During or after completing your 1989-90 degree, did you apply to any school for additional formal training? (yes/no)

12. Have you attended school at any time since receiving the 1989-90 degree? (yes/no)

QUESTION 13 WAS ONLY ASKED IF Q11 AND Q12 WERE BOTH ANSWERED NO

13. Which of the following best describes your reason for not applying to school? Would you say...

    You had no plans to continue your education,
    You wanted to work before continuing your education,
    You wanted to take time off before continuing your education, or
    You could not afford to continue your education?
    OTHER (SPECIFY)

For question 11, about 9 percent of the cases had discrepancies. For 40 percent of the discrepancies the reason cited was that the question was unclear. The time reference for this question, "During or after completing your 1989-90 degree," is somewhat complicated. Also, the term "formal training" may be ambiguous. In the other (specify) responses, some graduates indicated that they were unsure what types of training should be included in this question.

For question 12, about 9 percent of the cases had discrepancies. Among the discrepancies, the main reasons cited were that the situation had changed (30 percent), and the question was unclear (23 percent). It is not surprising that almost one-third of the discrepancies were caused by the situation changing, since graduates who began attending school between the original survey and the reinterview would be included in this category.

For question 13, about 26 percent of the cases had discrepancies. Among the discrepancies, the main reasons cited were that the graduate's perception had changed (24 percent), the question was unclear or the response category used in the question did not fit the graduate's situation (23 percent), and other reasons (20 percent). Since this is an opinion question, it is understandable that the graduates' perceptions would change. It is also understandable that some graduates would say that the response categories did not fit their situations, or that more than one answer was correct (the most common response for the other category).

Question 23. Only about 4 percent of the cases had discrepancies for this question on whether the graduate was working for pay the week of April 22, 1991. Over half (57 percent) of the discrepancies were reported as recall problems.

Question 32. About 6 percent of the cases had discrepancies for whether the job was full time or part time during the week of April 22, 1991. The most common reasons for the discrepancies were that the question was unclear (30 percent) and that the graduate's perception had changed (26 percent). Both of these reasons seem to indicate that some of the graduates were unsure of the definition of full and part time.

Question 34. This question was read to graduates as follows:

34. Were you:

An employee of a corporation, private company, business, or individual, for wages, salary, or commissions,
A federal government employee,
A state government employee,
A local government employee (city, county, etc.), or
Self-employed in your own business, professional practice or firm?

Less than 7 percent of the cases had discrepancies for this question. Of those with discrepancies, 41 percent said the question was unclear or the response category used in the question did not fit their situation, 30 percent said that it was difficult to recall an exact answer, and 22 percent said their perception had changed since the first interview was conducted.

Question 38. This question asked how many hours per week the graduate was usually employed on the principal job held the week of April 22, 1991. About 36 percent of the cases had discrepancies for this question. It should be noted that, since this question asks for the specific number of hours, even a difference of 1 hour per week would appear as a discrepancy. Almost half (47 percent) of the graduates with discrepancies said that it was difficult to recall an exact answer to the question.

Question 39. This question asked at what rate (before deductions) the graduate was paid on the principal job held the week of April 22, 1991. The question asks for the amount and the unit (year, month, week, day, or hour). About one quarter (24 percent) of the cases answered in a different unit on the reinterview than on the original survey. Of these cases, 43 percent gave an "other" reason for the discrepancy. Most of these other reasons were that the answers were the same but given in different units. An additional 30 percent of the cases that answered in different units said that it was difficult to recall an exact answer to the question.

About 34 percent of the cases gave their salary in the same unit but gave a different salary amount in the reinterview than in the original survey. Even small differences due to rounding would appear as discrepancies. Among the cases with discrepancies, 57 percent said that they had difficulty recalling an exact answer to the question.

Question 40. About 6 percent of the cases had discrepancies on this question, which asked whether, in addition to the principal job, the graduate was working for pay at a second job during the week of April 22, 1991. Over half (58 percent) of the cases with discrepancies said that they had difficulty recalling an exact answer to the question.

Question 53. Only 10 cases, or 2 percent, had discrepancies for this item, which asked whether the graduate was certified to teach. Of these, 4 respondents said the situation had changed.

Question 67. About 24 percent of the cases had discrepancies for this item, the month in which the graduate first started teaching. Over half (54 percent) of the cases with discrepancies said that they had difficulty recalling an exact answer to the question. An additional 27 percent said that the question was unclear.

APPENDIX D

MEASUREMENT ERRORS UNDER COMPLEX SAMPLES

APPENDIX D
MEASUREMENT ERRORS UNDER COMPLEX SAMPLES

This appendix provides some of the mathematical foundations supporting the use of the weighted measurement error statistics for complex sample designs. In particular, the net and gross difference rate and the index of inconsistency for more general sample designs are investigated. This development follows the same approach used by Hansen, Hurwitz and Bershad (1961) and later by Biemer and Stokes (1991). Before studying the measurement error statistics, the measurement error model is first introduced.

Measurement Error Model

The simplest model for measurement error in a sample survey assumes that the observed value at any interview (trial) can be written as the true value plus an additive error term:

$$y_{ti} = \mu_i + \varepsilon_{ti} , \qquad \text{(D.1)}$$

where μ_i is the true value for unit i and ε_{ti} is the error of observation at trial t.

Consider estimating a total (or mean or other simple linear statistic) under this measurement error model. The estimated total for a characteristic, y, is

$$y' = \sum^{n} w_i y_i \qquad \text{(D.2)}$$

where w_i is the sampling weight for unit i.

The expected value of the estimated total is found by first taking the expectation conditional on the model (E_2) and then taking the expectation over all possible samples (E_1). The expected value is

$$
\begin{aligned}
Ey' &= E_1E_2\sum^n w_i y_i \\
&= \sum^N E_1(\delta_i)w_iE_2(\mu_i+\varepsilon_{ti}) \\
&= \sum^N \mu_i + \sum^N \beta_i \quad , \\
&= \sum^N \mu_i + N\beta
\end{aligned} \tag{D.3}
$$

where $\delta_i = 1$ if unit i is included in the sample and $\delta_i = 0$ if not, $\beta_i = E_2\varepsilon_{ti}$ is the response bias for unit i and $\beta = \frac{1}{N}\Sigma^N\beta_i$. This derivation assumes that the weights are the inverse of the probabilities of selection of the units, i.e., that $E_1\delta_i = \frac{1}{w_i}$. If β is zero, as occurs when $\beta_i=0$ for all i, then the estimated total is unbiased, i.e., $Ey' = N\mu = \sum^N \mu_i$.

The total variance of the estimate is the sum of its sampling variance and its response variance. For the estimated total, the total variance of the estimate can be written as

$$
\begin{aligned}
V(y') &= V(\sum^n w_i(\mu_i + \varepsilon_{ti})) \\
&= V(\sum^n w_i\mu_i) + V(\sum^n w_i\varepsilon_{ti}) + Cov(\sum^n w_i\mu_i, \sum^n w_i\varepsilon_{ti}) \ , \\
&= \sigma^2_{\mu'} + V(\sum^n w_i\varepsilon_{ti}) + 0
\end{aligned} \tag{D.4}
$$

where $\mu' = \sum^n w_i\mu_i$ and the covariance term is zero under the condition that the errors are uncorrelated with the true population values.

The first term of the equation (D.4) is the sampling variance of the estimate (the squared standard error of the estimate) when the values are observed without measurement error. The second term is the response variance of the estimate.

The response variance of the estimate can be expressed as

$$
V(\sum^n w_i\varepsilon_{ti}) = \sum^n w_i^2 V(\varepsilon_{ti}) + \sum_{i \neq j}\sum w_i w_j Cov(\varepsilon_{ti}, \varepsilon_{tj}) \tag{D.5}
$$

The first term on the right hand side of (D.5) is the simple response variance of the estimate and the second term is the correlated component of the response variance of the estimate.

When the error terms are uncorrelated, i.e., $Cov(\varepsilon_{ti}, \varepsilon_{tj}) = 0$, the correlated component of the response variance vanishes. Theorem D.2 in Wolter (1985) shows that if this covariance is zero, a nearly unbiased estimate of the total variance of the estimate is given by applying standard variance estimation formula with the observed sample values (which are collected with measurement error). The rest of our discussion will be based on this simplifying assumption. Since the correlated component is often the most significant factor in the measurement error, further research is needed when this assumption is eliminated.

Response Bias

The first measurement error considered is the response bias and how it can be estimated from the reinterview data. Under a sampling scheme in which estimation weights (w_i) are attached to each sampled unit, the net difference rate can be written as

$$ndr = \frac{\sum^n w_i(y_{1i} - y_{2i})}{\sum^n w_i} \tag{D.6}$$

where y_{1i} is the observed value for unit i in the original survey and y_{2i} is the observed value for the same unit in the reinterview. Under simple random sampling, this reduces to the usual estimator for the net difference rate,

$$ndr = \frac{\sum^n (y_{1i} - y_{2i})}{n} . \tag{D.7}$$

Now, the expected value of the net difference rate can be evaluated using the measurement model given in (D.1). As before, the expected value is found by first taking the expectation conditional on the model (E_2) and then taking the expectation over all possible samples (E_1).

$$\begin{aligned} E(ndr) &= E_1 \frac{1}{\sum^n w_i} E_2\{\sum^N \delta_i w_i (y_{1i} - y_{2i})\} \\ &= E_i \frac{1}{\sum^n w_i} E_2\{\sum^N \delta_i w_i (u_i + \varepsilon_{1i}) - \delta_i w_i (u_i + \varepsilon_{2i})\} \\ &= E_i \frac{1}{\sum^n w_i} \{\sum^N \delta_i w_i E_2(\varepsilon_{1i} - \varepsilon_{2i})\} \\ &= E_i \frac{1}{\sum^n w_i} \{\sum^N \delta_i w_i (\beta_{1i} - \beta_{2i})\} \end{aligned} \tag{D.8}$$

Equation (D.8) shows that the net difference rate has an expected value of zero if $\beta_{1i} = \beta_{2i}$. Notice that under this condition, the expectation is zero for both the weighted and unweighted net difference rate. This condition holds under a model which assumes the distribution of the errors is the same from trial to trial.

Studies of response variance based on reinterviews attempt to simulate the conditions where the error terms are identically distributed (the trials are independent and conducted under the same general conditions). Of course, the ability to do this is limited by any conditioning effects, i.e., the possibility that respondents' answers to the second interview are affected by the fact that they had been interviewed before. The conditioning effects on the respondents are assumed to be negligible for this research, but this is often a questionable assumption.

In the RCG, the first interview and the reinterview constitute a response variance type of study. The general conditions in the interview and the reinterview are very similar (except the omission of a few items from the reinterview). In this setting, the net difference rate should have an expected value of zero, even if the results are not weighted. Estimates of net difference rates (based on the original interview and the reinterview, not the reconciled values) that are significantly different from zero are indications that the assumptions of a response variance study are not being met.

If the error terms across interviews do not have the same distribution, then the unweighted analysis is not unbiased for the population net difference rate ($N\beta_1 = N\beta_2$). The net difference rate is unbiased if the weights are inversely proportional to the probabilities of selection of the units (the same condition imposed in the derivation of equation (D.3)) and the sum of the weights is a constant. The last condition is met under several designs, e.g., when all of the weights are constant and the total sample size is fixed, or when the sample is poststratified to a known total. Under the RCG, and most other complex sample designs, these conditions are not met exactly, but the approximation is often reasonable.

If the conditions noted above hold, then the expectation of the net difference rate is given by

$$\begin{aligned} E(ndr) &= E_1 \frac{1}{\sum^n w_i} \{\sum^N \delta_i w_i E_2(\varepsilon_{1i} - \varepsilon_{2i})\} \\ &= \frac{1}{N}\{\sum^N E_i(\delta_i) w_i (\beta_{1i} - \beta_{2i})\} \\ &= \frac{1}{N}\{\sum^N (\beta_{1i} - \beta_{2i})\} \\ &= \beta_1 - \beta_2 \end{aligned} \qquad \text{(D.9)}$$

Response bias studies attempt to simulate conditions where the latter trial is a 'better' measure of the true value by using more highly trained interviewers or probing techniques. In such studies, it is common to assume that the second trial is conducted without error ($\varepsilon_{2i} = 0$) and then the net difference rate (using the weighted analysis) is an unbiased estimate of the response bias in the estimate from the first interview.

In the RCG, the reconciliation of the first and second interviews is an attempt to develop a measure with little or no measurement error that satisfies the conditions for a response bias study. Assuming the reconciliation is the true value, the net difference rate computed with y_{1i} the response in the original interview and y_{2i} the response from the reconciled reinterview provides an estimate of the response bias of the estimate when the weights are used in computing the rate.

Simple Response Variance

Following the original development given by Hansen, et al.(1961) for simple random sampling, the gross difference rate and its relationship to measurement error is now examined. First, the definition of the gross difference rate is extended to complex samples in the same way as for the net difference rate. The gross difference rate can be written as

$$\begin{aligned} gdr &= \frac{1}{\sum^n w_i} \sum^n w_i (y_{1i} - y_{2i})^2 \\ &= \frac{1}{\sum^n w_i} \sum^n w_i (\mu_i + \varepsilon_{1i} - u_{i'} - \varepsilon_{2i})^2 . \\ &= \frac{1}{\sum^n w_i} \sum^n w_i (\varepsilon_{1i} - \varepsilon_{2i})^2 \end{aligned} \qquad \text{(D.10)}$$

As with the net difference rate, this expression reduces to the ordinary expression for the gross difference rate under simple random sampling.

If the conditions for a valid response variance study are satisfied (i.e., the first and second moments of the error terms are identical and the errors between trials are uncorrelated), then expectation of the gross difference rate is directly related to the simple response variance. This follows from

$$E(gdr) = E_1 E_2 \frac{1}{\sum^n w_i} \sum^n w_i (\varepsilon_{1i} - \varepsilon_{2i})^2$$
$$= E_i \frac{1}{\sum^n w_i} \sum^N \delta_i w_i E_2 (\varepsilon_{1i}^2 + \varepsilon_{2i}^2 - 2\varepsilon_{1i}\varepsilon_{2i})^2, \tag{D.11}$$
$$= \frac{2}{N} \sum^N E_1(\delta_i) w_i \sigma_{\varepsilon i}^2$$

where $\sigma_{\varepsilon i}^2 = E_2(\varepsilon_{1i}^2) = E_2(\varepsilon_{2i}^2)$ is the variance of the error term for unit i.

Defining the population simple response variance as $\sigma_\varepsilon^2 = \frac{1}{N}\sum^N \sigma_{\varepsilon i}^2$, the expected value of the gross difference rate is:

$$E(gdr) = \frac{2}{N}\sum^N \sigma_{\varepsilon i}^2 . \tag{D.12}$$
$$= 2\sigma_\varepsilon^2$$

Therefore, if the weight is inversely proportional to the probability of selection of the unit, the sum of the weights is a constant and the conditions for a response variance study are satisfied, then the weighted gross difference rate is an unbiased estimate of twice the simple response variance.

While the results stated above on the gross and net difference rates match the simple random sampling results, these parallels do not extend to all estimates. For example, under simple random sampling and the assumptions noted above, it is easy to show that the gross difference rate divided by the sample size is an approximately unbiased estimator of the variance of the net difference rate. This result does not apply in more complex sampling schemes. A self-weighting scheme is a sufficient condition for this result to hold under other sample designs. The RCG design is not self-weighting.

Relative Impact of Measurement Error

The last measurement error statistic of interest is the index of inconsistency. The index of inconsistency is normally defined as the ratio of the simple response variance to the total variance, assuming again that there are no correlated response errors. The index of inconsistency is defined in terms of population variances, not variances of the estimates. In general, the index can be written as

$$I = \frac{\text{simple response variance}}{\text{total variance}} = \frac{\sigma_\varepsilon^2}{\sigma_\mu^2 + \sigma_\varepsilon^2} \quad \text{(D.13)}$$

For example, consider the case of simple random sampling. The unweighted gross difference rate divided by two is an unbiased estimate of the simple response variance (σ_e^2) and is the numerator of the index. The total variance is estimated by an unbiased estimate of the population variance, normally the square sample standard deviation. For dichotomous variables, the total variance is estimated by p(1-p), where p is the sample proportion. This is the index of inconsistency for simple random samples with no correlated response variance, as described by Hansen, et al.

For more complex sampling and estimation schemes, a consistent estimator for the total variance can be used in the estimation of the denominator of the index. Following Kish (1965), page 68, a consistent estimate of the variance can be written as

$$v_y = \frac{\sum^n w_i (y_i - \tilde{y})^2}{\sum^n w_i} , \quad \text{(D.14)}$$

where $\tilde{y} = \frac{\sum^n w_i y_i}{\sum^n w_i}$ is the weighted sample mean. As Kish shows, this estimate is biased and can be made unbiased by adding $V(\tilde{y})$. Since this term is neglible compared to v_y whenever the sample size is relatively large, it can be ignored for estimating the total variance in the RCG and most other large sample surveys.

For dichotomous variables, v_y reduces to the binomial variance formula using the weighted estimates of the proportion. In this case, the index can be estimated as

$$I = \frac{\text{gdr}}{2\hat{p}(1-\hat{p})} , \quad \text{(D.15)}$$

where $\hat{p}$ is the correctly weighted estimate of the proportion of the population in the category and the gross difference rate is given by equation (D.10). This was the approach used in the RCG.

Other options are available for estimating the denominator of the index of inconsistency. For example, the estimated proportion could be estimated based on the reinterview sample only or could be a combined estimate of the proportion from the reinterview and the full sample. These options are analogous to those available for the simple random sampling index.

APPENDIX E

STATE CERTIFICATION AGENCY SURVEY FORM

US DEPARTMENT OF EDUCATION
Evaluation Study of Teacher Certification - Graduate Form

Please provide the following information for the graduate listed on the label above, and return this form to Westat. If you have any questions, please call Cindy Gray of Westat at 1-800-937-8281.

1. Does the graduate currently hold any type of regular or temporary teaching certificate or licence to teach school at any grade level(s), prekindergarten through grade 12, in your state?

Yes	1	(SKIP TO QUESTION 3)
No	2	(GO TO QUESTION 2)

2. At any time during 1991, did the graduate hold any type of regular or temporary teaching certificate or licence to teach school at any grade level(s), prekindergarten through grade 12, in your state?

Yes	1	(GO TO QUESTION 3)
No	2	(SKIP TO QUESTION 7)
Not able to determine................	8	(SKIP TO QUESTION 7)

3. What kind of certificate or license does/did the graduate have? (CIRCLE ONE)

Initial or provisional certificate leading to regular or standard certificate	1
Regular or standard certificate	2
Alternative, emergency, or temporary certificate	3
OTHER (SPECIFY)________________________............	4

INSTRUCTIONS FOR ANSWERING QUESTIONS 4 - 6:
In answering questions 4 through 6, please include any kind of certificate or license to teach school at any grade level prekindergarten through grade 12 (including regular, provisional, alternative, emergency, and temporary certificates).

4. In what month and year did the graduate first receive a certificate or license to teach?

MONTH:__________ YEAR:__________

5. In what grades is/was the graduate certified to teach? (*CIRCLE ALL THAT APPLY*)

PREKINDERGARTEN	P	SEVENTH	7
KINDERGARTEN	K	EIGHTH	8
FIRST	1	NINTH.............................	9
SECOND	2	TENTH	10
THIRD	3	ELEVENTH	11
FOURTH..............................	4	TWELFTH	12
FIFTH	5	UNGRADED	13
SIXTH	6	ALL GRADES	14
		SUBJECT CERTIFIED	15

(CONTINUED, OVER)

6. Please circle below the fields in which the graduate has/had specific subject certification to teach: (*CIRCLE ALL THAT APPLY*)

		YES--CERTIFIED
1.	Any Elementary fields, general or specialized	1
2.	Art/fine art/performing arts	1
3.	Basic skills and remedial education	1
4.	Bilingual education	1
5.	Biological or life sciences	1
6.	Business (not part of voc. ed. curriculum)	1
7.	Computer science	1
8.	English language arts	1
9.	English-as-a-second language	1
10.	Foreign languages	1
11.	Gifted/talented	1
12.	Health	1
13.	Home economics	1
14.	Industrial Arts, Trade, and Industry	1
15.	Mathematics	1
16.	Music	1
Any Physical sciences, general or specialized:		
17.	General Sciences (no specialized area)	1
18.	Chemistry	1
19.	Geology/earth science	1
20.	Physics	1
21.	Other physical sciences	1
22.	Physical education	1
23.	Pre-elementary education	1
24.	Reading	1
25.	Religion/philosophy	1
26.	Social science/social studies	1
Any Special education fields:		
27.	Mentally retarded	1
28.	Hearing impaired, deaf	1
29.	Seriously emotionally disturbed	1
30.	Speech impaired	1
31.	Specific learning disability	1
32.	General certificate (no specific condition)	1
33.	Other special education	1
34.	Vocational Education, other than Business, Home Economics, or Industrial Arts	1
35.	Other fields	1

7. Please provide any additional information that might help us understand this graduate's certification:

____

____

____

____

APPENDIX F

CERTIFICATION SURVEY CODING RULES

APPENDIX F

CERTIFICATION SURVEY CODING RULES

RECENT COLLEGE GRADUATES
State Certification Form Coding Instructions

Questions A-D involve overall comparisons of the graduate responses against the state reported data. Questions E-F are coded with specific grade and field information, as reported by the state. However, some comparison with graduate information will be necessary in assigning the code "3" to questions E-F.

It seems easiest to code the grade questions first, then the field questions. In addition, it will be easier to code the specific fields (in part F) before coding the overall fields (part D). Therefore, the instructions are listed in this order, rather than in the order they appear on the form.

On the state form, Q7 was included to help clarify the certification information. You should always read this information and use it to help code the state-reported data.

A. Certification confirmed:

All graduates in this study reported that they were certified. If the state reports that the graduate is certified, then the certification is confirmed.

1. Code QA as "yes" if the state answers "yes" to Q1 or Q2, or otherwise indicates that the graduate is certified. AR did not answer Q1 or Q2 on any of their forms, but entered all of the certification information on the form - these cases should be coded as "1," not nonresponse.

2. Code QA as "no" if the state answers "no" to Q1 and Q2. The rest of the coding form will be left blank.

3. The category of "Yes, but reported as statement of eligibility by state" is used for Florida only. In Florida there are several cases where Q1 is answered no, but Q7 indicates that the graduate has a statement of eligibility. Pull these cases and any that report the graduate has a substitute certificate.

4. Code QA as "State not able to determine whether certified" if Q2 is answered "Not able to determine." The rest of the coding form will be left blank.

B. Kind of Certification:

The graduate form lists the kind of certification in field Q56. Four codes are possible (in the same order as the State form):

1 = Initial or provisional certificate leading to regular or standard certificate
2 = Regular or standard certificate
3 = Alternative, emergency, or temporary certificate
91 = Other (specify)

Circle the appropriate code in QB by matching the kind reported in Q56 by the graduate to the kind reported in Q3 by the state. The order of priority for coding "2" and "3," from highest to lowest, is:

Regular or standard
Initial or provisional
Alternative, emergency, temporary

C. Grades certified to teach (overall):

Circle the appropriate code in part C, and then code the specific grade information in part E.

E. Specific grades certified to teach (as reported by state):

1. If the state reports "yes" for a grade, code that grade as "1."

2. If the state reports "all grades," then code all the grade categories except subject certified as code "1." (In CA, almost all the state forms indicate "all grades").

3. If the graduate reports a grade not reported by the state, look to see whether the grade might be confirmed by the state but reported in a different way. Follow these rules for using code "3" (some examples of coding/reporting differences are on the attached page):

   a. If the graduate reports "all grades" and the state reports K-12 or 1-12, then code the grades reported by the state as code "1" and code the rest of the grades (except subject certified) as code "3."

   b. For any other situation where it appears that the state confirms the graduate information but coding/reporting differences exist, code the grade(s) as "3" or make a problem sheet for supervisor review.

4. If the state and graduate both report "no" to a grade, code that grade as "2."

5. If the graduate reports a grade not reported by the state and no coding/reporting differences exist, code the grade as "2."

Subject certified and grades given, code 4.

F. Specific fields certified to teach (as reported by state):

1. In Michigan, none of the state forms have "Elementary" circled in the list of fields. Therefore, we will assume that forms with grades K-8 are elementary.

2. If the state reports "yes" for a field, code that field as "1."

3. If the graduate reports a field not reported by the state, look to see whether the field might be confirmed by the state but reported in a different way. Follow these rules for using code "3" (some examples of coding/reporting differences are on the attached page):

   a. If both the state and graduate report elementary certification, but the graduate has other specific subjects (such as Basic skills, English, science, math, reading, social studies, etc.), code the specific subjects as "3," UNLESS the specific subject is special education.

   b. Code elementary as "3" on the coding form:
   If the graduate has code "3" for elementary (a code that we assigned because the graduate is certified in at least one grade K-5), and
   If the state has elementary answered as "no."

   c. Code elementary as "3" on the coding form:
   If the graduate has code "1" for elementary and
   The state has elementary answered as "no," and
   The state shows the graduate is certified in at least one grade K-5 (or "all grades").

   d. If both the graduate and the state report certification in special education, but report it in different fields, then:
   Code the special education fields reported by the state as "1"
   Code the special education fields reported by the graduate but not by the state as "3."

   e. For any other situation where it appears that the state confirms the graduate information but coding/reporting differences exist, code the field(s) as "3" or make a problem sheet for supervisor review.

4. If the state and graduate both report "no" to a field, code that field as "2."

5. If the graduate reports a field not reported by the state and no coding/reporting differences exist, code the field as "2."

D. Subjects certified to teach (overall):

Code 1: If all the same subjects were reported by graduate and state.

Code 2: If the graduate reported elementary and specific subjects, state confirms elementary. (Part F has elementary = 1 or 3 and other subjects = 3)

Code 3: If graduate reported elementary (as instructed because teaching elementary grades) and one or two specific subjects (such as phys ed, health, art, music). State confirms subject(s) but not elementary.
(Part F has elementary = 3 and one or two specific subjects, such as phys ed, health, art, music, foreign language, reading = 1)

Code 4: If State confirms special education certificate but chooses different specific categories (such as "general" or "other")
(Part F has at least one special education field = 3)

Code 5: Some subjects confirmed, some not confirmed
(Some subjects reported by the **graduate** are coded 1 or 3 in part F, and some are coded 2 in part F)

Code 6: None of the subjects confirmed
(All of the subjects reported by the **graduate** are coded 2 in part F)

Code 9: Nonresponse (by graduate or state)

EXAMPLES OF CODING/REPORTING DIFFERENCES

Specific grades certified to teach:

- For grades, the main coding/reporting differences involve the categories of: prekindergarten, ungraded, all grades, and subject certified. During the survey data collection, if the graduate reported certification in "all grades," the "all grades" category was coded by the interviewer and the CATI system automatically coded all other categories in the question except subject certified. If a respondent considered certification in K-12 to be "all grades," then certification in Pre-K and "ungraded" may not be confirmed by the state.

- The "ungraded" category was meant mainly to capture special education where certification is often given by ages, rather than by grades. However, since no rules were set in the survey or on the state forms for use of this category, it is subject to interpretation. Some respondents report the grades that correspond to the ages certified to teach, some report "ungraded," and some report "subject certified."

- On the survey, the "subject certified" category was meant for those people who were not certified by grade, but **only** by subject. Interviewers were instructed to probe for grades in which the respondent was certified to teach a specific subject, and only code "subject certified" if a respondent confirmed that he/she was not certified by grade. If "subject certified" was coded, then no other grade could be coded. However, some of the states have circled specific grades and "subject certified" to indicate that the graduate was certified by grade and subject.

Specific fields certified to teach:

- For the certification fields, the main coding/reporting differences involve elementary and special education certification. During the survey data collection, the category "Any elementary fields, general or specialized" was meant to include any respondent certified to teach any subject at the elementary level. Respondents were then expected to answer yes to the specific subject fields only if they had an additional certification in that field. In practice, however, some respondents answered yes to each of the subject fields included in their elementary certification, rather than only those in addition to elementary. This may have been exacerbated by the fact that we read each category to the respondents during data collection. However, most of the state forms were completed according to the original intent for elementary certification -- that is, the specific subject fields were circled only if the graduate had an additional certification in that field.

- The second problem that occurs with elementary certification involves graduates certified to teach elementary grades but only in a specialized subject (such as phys ed, art, music, reading). On the survey, these graduates reported "yes" to "Any elementary fields," since they were certified on the elementary level. However, most of the states did not consider this to be elementary certification.

- For graduates certified in special education, there can be different interpretations of how to fit the certification into the survey categories. For example, a graduate certified to teach "Mildly handicapped K-12," answered yes to the specific handicapping conditions that the certificate covers (such as mentally retarded and specific learning disability). However, the state chose the category "General certificate, no specific condition," presumably since no specific condition is named in the certificate. Again, the different interpretations may have been increased since graduates were asked whether they were certified in each of the specific special education fields.

APPENDIX G

STATE-BY-STATE ANALYSIS OF REPORTING DIFFERENCES FOR KIND OF CERTIFICATE

APPENDIX G

STATE-BY-STATE ANALYSIS OF REPORTING DIFFERENCES FOR KIND OF CERTIFICATION

Both the graduate and state were asked to choose one of the following categories for kind of certification:

- Initial or provisional certificate leading to regular or standard certificate;
- Regular or standard;
- Alternative, emergency, or temporary certificate; and
- Other (specify).

The main reason for reporting differences in the kind of certificate appears to be different interpretations of the reporting categories. None of the 10 states included in the validity study use classifications exactly the same as those used on the survey. By looking at the classifications used in each state and the response patterns for that state, explanations for the reporting differences often emerge. The match rates for each state appear in the table below, and are discussed in the following sections.

Percentage of cases with kind of certificate reported the same, gross difference rates, and net difference rates, by state

State agency	Sample size	Match on kind of certificate		
		Percent reported the same	Gross difference rate	Net difference rate
Total	306	57.5	42.5	9.2
Arkansas	30	86.7	13.3	6.7
California	26	65.4	34.6	-11.5
Florida	26	53.8	46.2	-30.8
Indiana	24	79.2	20.8	20.8
Michigan	30	96.7	3.3	-3.3
Ohio	28	21.4	78.6	78.6
Pennsylvania	46	30.4	69.6	39.1
Tennessee	30	53.3	46.7	-46.7
Texas	43	55.8	44.2	44.2
Utah	23	47.8	52.2	-52.2

Arkansas. About 87 percent (26 of 30) of the cases in Arkansas were classified the same by both the graduate and the state. Most of these matched cases (24) were classified by both the graduate and the state into the RCG category of "regular or standard." Of the 4 cases that were classified differently by the graduate and the state, 3 were identified by the graduate as "initial or provisional" and by the state as "regular or standard."

Arkansas has a 6-year certificate for bachelor's degree recipients and at least six different provisional certificates. One possible area of confusion for graduates is that the Arkansas application materials refer to "initial certification" to identify those applying for the standard 6-year certificate for the first time. Thus, those who apply for and obtain this "initial certification" are actually obtaining a "regular or standard certificate."

A second possible reason for differences in the graduate and state reported data is the different data collection time periods. One type of provisional certificate is given to applicants who meet all other requirements except having an acceptable score on the National Teacher Examination (NTE). Graduates who were given a provisional certificate and then took the examination may have changed from a provisional to a standard certificate during the time between data collections.

California. About 65 percent (17 of 26) of the cases in California had certification type reported the same by both the graduate and the state. California has two types of teaching credentials: (1) a Multiple Subject Teaching Credential that authorizes the holder to teach in a self-contained classroom such as the classrooms in most elementary schools; and (2) a Single Subject Teaching Credential that authorizes the holder to teach the specific subject(s) named on the credential in departmentalized classes such as those in most junior high and high schools. For each of these credential types, there are three levels:

- One-Year Preliminary Credential. This may be obtained with a bachelor's degree or higher, completion of a teacher preparation program, and passage of the California Basic Educational Skills Test.

- Five-Year Preliminary Credential. This is obtained through a 4-year extension to the first credential, which requires minimum scores to certain sections of the National Teacher Examination or additional course work.

- Professional Clear Credential. This requires completion of a fifth year of study after the bachelor's degree and completion of courses in specific areas.

Since the California certification levels are not identified using the same terminology used in the RCG survey, it is not clear how each level of credential should fit into the RCG categories. A 1-year or 5-year preliminary credential might be interpreted as "initial or provisional," "regular or standard," or "temporary." Of the nine cases that were not matched, four were categorized as "regular or standard" by the graduate and "initial or provisional" by the state. Another four of the unmatched cases were categorized as "alternative, emergency, or temporary certificate" by either the graduate or the state, but not by both.

Florida. About 54 percent (14 of 26) of the cases in Florida had certification type reported the same by both the graduate and the state. The Florida Department of Education identifies three steps or levels in the certification process:

- Statement of Eligibility. Statutes and rules that govern the issuance of Florida Educator's Certificates require that the individual be employed in a public or private elementary or secondary school with an approved Professional Orientation Program before a certificate is issued. Applicants are, therefore, provided a Statement of Eligibility for use in obtaining employment.

- Two-Year Nonrenewable Temporary Certificate. This certificate may be obtained by those who hold a valid statement of eligibility, are employed in a school with an approved Professional Orientation Program, and have submitted fingerprints.

- Five-Year Professional Certificate. This certificate is issued to those who meet the requirements for the Temporary Certificate, satisfy the coursework and test score requirements, and have completed the Professional Orientation Program.

Of the 12 unmatched cases, 8 were classified as "alternative, emergency, or temporary certificate" by the graduate and as "initial or provisional" by the state. It seems likely that both the graduate and the state were referring to the 2-year nonrenewable Temporary Certificate for these cases, but chose to classify it into different RCG categories. In fact, the cover letter sent from the state certification agency refers to the Temporary Certificate as the initial certificate. However, it is understandable that the word "temporary" in the name of the certificate caused graduates to choose the RCG category that contained that word.

Indiana. About 79 percent (19 of 24) of the cases in Indiana had certification type reported the same by both the graduate and the state. The Indiana Department of Education describes the following three types of certificates:

- Standard License. Applicants who meet all of Indiana's certification requirements in their licensing area(s), including the teacher competency tests and recency credit, are eligible for an Indiana Standard License. The Standard License is valid for 5 years and maybe renewed indefinitely by completing six semester hours of approved credit every 5 years.

- Reciprocal License. Out-of-state graduates who do not meet all of Indiana's certification requirements but hold an unexpired out-of-state license may be eligible for a 1-year Reciprocal License. The Reciprocal License may be renewed up to four times by completing necessary tests and course work.

- Professional License. Applicants who meet the requirements for the Standard License and who have completed a master's degree with appropriate course work and have 5 years of teaching experience in an accredited school may be eligible for a Professional License. The Professional License is valid initially for 10 years, then renewable every 5 years on the completion of 6 semester hours of approved academic credit.

All five of the unmatched cases were classified as "initial or provisional" by the graduate and as "regular or standard" by the state. All had graduated from an Indiana school. It appears that Indiana uses the word "initial" to refer to the first time an individual obtains certification. These individuals must complete a teacher internship program, as described in the certification brochure:

> Individuals receiving an initial Standard or Reciprocal teaching license will be required (Public Law 390 - 1987) to successfully complete a one-year (two semester) beginning teacher internship. Individuals with two (2) years teaching experience in an accredited out-of-state school will not be required to complete the internship. Do not be concerned about the internship program until you receive your initial Standard or Reciprocal License and are employed in an accredited Indiana school. At that point, consult your principal and/or superintendent for details.

The five graduates who reported their certification as "initial or provisional" may have considered their certification to be initial until they completed the internship program. Since the state uses the term "initial Standard" to describe the first time an individual obtains a Standard License, graduates might reasonably have chosen either RCG category "initial" or "standard."

Michigan. Michigan had the highest rate of matching on certificate type of all the states in the survey. Of the 30 cases in the state, 29 were matches (97 percent). The teacher certification brochure produced by the Michigan State Board of Education describes the types of certificates in that state as follows:

> There are four basic types of Michigan regular and vocational certificates currently available: the required initial certificate, called the Provisional certificate; the Continuing certificate, which may eventually be obtained when the holder of a Provisional certificate meets the requirements as outlined in the "Continuing Certificate Requirements" section of this brochure; the Temporary Vocational Authorization; and the Full Vocational Authorization.

Most cases (24 of 30) were categorized by both the graduate and the state in the RCG category "initial or provisional." Apparently, the state's use of the terms "initial" and "provisional" to describe the first level of certification made it easy for the graduates and state agency to choose the same RCG category. In fact, only one case was categorized differently by the graduate and the state. For this case, the graduate chose "regular or standard" and the state chose "initial or provisional." The graduate may have been confused by the term "regular," which is used by Michigan to differentiate their non-vocational certificate from their vocational certificate.

Ohio. Ohio had the lowest match rate for certification type with 6 of 28 cases (21 percent) matching. The Ohio certification levels are described by the Ohio Department of Education as follows:

> Initial standard Ohio certificates are called provisionals and are valid for four years. Regardless of the grade of certificate you may currently hold in Ohio or in any other state, the initial certificate will be issued as a four-year provisional. Provisional certificates may later be converted to professional and then to permanent certification.

All except one of the unmatched cases (21 of 22) were classified as "initial or provisional" by the graduate and as "regular or standard" by the state. It seems likely that both the graduates and the state were referring to the same certificate (the initial standard), since that is the most likely certificate for new graduates. The use of all three words -- initial, standard, and provisional -- to describe the first level of certification meant that either RCG category "initial or provisional" or "regular or standard" could have been chosen. However, in this case, the category chosen by most of the graduates (initial or provisional) seems more appropriate than the category chosen by the state (regular or standard).

Pennsylvania. About 30 percent (14 of 46) of the cases in Pennsylvania had certification type reported the same by both the graduate and the state. The categories of instructional certificates issued by Pennsylvania are the following:

- Instructional Level I Certificate (Provisional). Valid for 6 years of service. May be converted to Level II after 3 years of service on Level I; must be converted after 6 years of service on Level I.

- Instructional Level II Certificate (Permanent). Valid for the life of the holder. Requirements: 3 years of satisfactory teaching in Pennsylvania on the Level I certificate and completion of 24 semester hours of postbaccalaureate study.

- Intern Certificates. Valid for 3 calendar years. Requirements: a bachelor's degree without a teacher certification program; acceptance into and recommendation by a Pennsylvania college with an approved Intern program.

Most of the unmatched cases (25 of 32) were classified as "initial or provisional" by the graduate and as "regular or standard" by the state. It seems likely that both the graduates and the state were referring to the Level I certificate, the most common certificate for new graduates. It is not clear why the state would choose to classify the Level I certificate as "regular or standard," perhaps because it is the expected or "regular" certificate for new graduates, or perhaps because the instructional certificates are considered "regular" compared to the vocational certificates.

Tennessee. In Tennessee, 53 percent (16 of 30) of the cases were classified in the same category by both the graduate and the state. Tennessee has several different types of teaching licenses, as described below:

- Probationary Licenses. Initial 1-year license issued to applicants on the basis of completion of a bachelor's degree and an approved teacher education program and submittal of minimum qualifying scores on the NTE. Renewable. Successful completion leads to appropriate Apprentice-level license.

- Apprentice Licenses. Three-year license based upon satisfactory completion of the probationary year. Renewable. Successful completion leads to appropriate professional license.

- Teacher's Professional License. A 10-year license issued on the basis of satisfactory completion of the 3-year apprenticeship.

- Career Ladder Certificates (optional) - Career Levels I, II, and III. Ten-year certificates issued to applicants who voluntarily elect to be evaluated for these levels on the Career Ladder.

- Interim Probationary Licenses:

  - Type A. One-year license based on a minimum of a bachelor's degree and 6 quarter hours of professional education college credit. Renewable four times. Requires superintendent's intent to employ.

  - Type B. One-year license issued to applicants who meet all certification requirements but lack minimum qualifying scores on the NTE Core Battery or Specialty Area Test. Renewable one time. Requires superintendent's intent to employ.

  - Type C. Requires bachelor's degree, completion of preservice portion of an approved alternative prep program, statement of intent to hire from Tennessee Superintendent.

Of the 14 unmatched cases, 10 were classified as "regular or standard" by the graduate and as "initial or provisional" by the state. The remaining 4 cases were classified as "alternative, emergency, or temporary" by the graduate and as "initial or provisional" by the state. With the large number of different licenses issued in Tennessee (none of which use the exact terminology used in the RCG survey), it is understandable that many graduates chose different categories than the state and other graduates. In fact, it is not clear which of the RCG categories would best describe each of the Tennessee licenses. In addition, the application materials that we received from Tennessee do not list or describe the various certification levels. Therefore, graduates may not have been aware of exactly which certificate they had or what the possible certificates are for the state.

Texas. In Texas, 56 percent (24 of 43) of the cases were classified in the same category by both the graduate and the state. The certificates issued by Texas include the following:

- Provisional Certificate. Issued on the basis of completion of a BA degree from an approved teacher education institute, and satisfactory performance on comprehensive exams. Valid for life of holder.

- Professional Certificate (not required). Issued on the basis of completion of a BA degree, at least 30 additional graduate level hours in an approved graduate teacher education program, and 3 years of acceptable teaching experience. Valid for life.

- One-Year Certificate. Issued to an individual who possesses a standard out-of-state teacher certificate. If the Texas Education Agency determines by evaluation that the applicant satisfies all requirements for Texas certification except for the testing requirement(s), he/she may request issuance of a One-Year Certificate. The testing requirement must be met during the validity period of the One-Year Certificate to qualify for continued certification in Texas.

All of the 19 unmatched cases were classified as "initial or provisional" by the graduate and as "regular or standard" by the state. It is easy to see how the Texas Provisional Certificate could be classified in either RCG category. The use of the term "provisional" would indicate that it belongs in the first RCG category. However, the full description of the RCG category is "Initial or provisional certificate leading to regular or standard certificate." The Texas Provisional Certificate does not lead to a regular or standard certificate, but rather is valid for the life of the holder. The Texas Professional Certificate is optional, not required. Therefore, the Provisional Certificate could be considered the "regular or standard certificate."

Utah. In Utah, 48 percent (11 of 23) of the cases were classified in the same category by both the graduate and the state. The certificates issued by Utah include the following:

- Basic Certificate. Requires completion of bachelor's degree and approved teacher education program. Valid for 4 years.
- Standard Certificate. Same requirements as Basic Certificate plus 2 years of successful teaching experience during first 4 years of teaching and recommendation of employing school district. Renewable.

All 12 of the unmatched cases were classified as "regular or standard" by the graduate and "initial or provisional" by the state. It is likely that both the graduates and the state were referring to the Basic Certificate, since this is the most common certificate for new graduates. It is easy to understand the state's classification of "initial or provisional," since the Basic Certificate leads to the Standard Certificate. However, graduates may not be as familiar with the Utah certification process. In fact, the copy of the application materials that we obtained from the Utah State Office of Education do not include any reference to the certification levels or types. Graduates may only know that they applied for and obtained a state certification, and may assume that it is a "regular or standard certificate."

APPENDIX H

SUGGESTED QUESTIONNAIRE REVISIONS FOR TEACHER ELIGIBILITY AND CERTIFICATION

APPENDIX H

SUGGESTED QUESTIONNAIRE REVISIONS FOR TEACHER ELIGIBLITY AND CERTIFICATION

Questions 50, 51, 52, 58 - Eligible to Teach

Question 50, which asks whether the graduate is eligible to teach, was worded on the survey as follows: "Are you eligible to teach school at any grade level from prekindergarten through grade 12? That is, have you completed all coursework, including student or practice teaching, required for a regular or standard certificate or license to teach at any or all levels in at least one state?"

Change "Eligible To Teach" to "Eligible To Be Certified." One difficulty with the above definition is the term "eligible to teach," which is used throughout the section to collect information on eligibility by grade and subject field. A more precise term would have been "eligible for regular or standard certification." While this did not appear to cause problems with question 50, where the term is immediately followed by the definition, it did cause problems with question 58. In this question, graduates were asked to report the subject fields in which they were eligible to teach. Some graduates assumed (incorrectly) that if they were certified in a subject they must be eligible to teach in that subject. Others, especially substitute teachers, thought that if they were allowed by the school district to teach in a subject field, they must be eligible to teach in that field.

Eliminate Restriction of "Regular or Standard." Another difficulty with this "eligible to teach" definition is that it refers to coursework required for a regular or standard certificate. As discussed in Chapter 5, the difference between "initial or provisional" and "regular or standard" certification is very ambiguous in some states. This is one of the reasons that only 58 percent of the cases in the validity study sample had the type of certification reported the same by both the graduate and state. Initial or provisional certification is quite common among new graduates, with about 29 percent of the certified graduates reporting this category on the main survey. A number of states require graduates to obtain an initial or provisional certificate and fulfill certain requirements (such as teaching for a specified time or completing an in-service course) before they can apply for a regular or standard certificate. For these reasons, it does not seem appropriate to limit eligibility to only regular or standard certification.

Collect Eligibility Only for Grades and Subjects in Which the Graduate Is Not Certified. If the definition of eligible were changed to no longer be restricted to regular or standard certification, then all certified graduates would be eligible by definition. This change would still allow the same type of analysis of certification and eligibility data that has been done in the past. For analysis, the subject eligibility and certification data were used to compare to the subject fields in which the graduate was teaching. The following three categories have been used for this analysis: (1) eligible or certified in some field; (2) eligible or certified in teaching field; and (3) certified in teaching field. Thus, graduates who are certified are included in the "eligible or certified" group, regardless of whether or not they are eligible. Therefore to conduct this analysis, it is necessary to determine eligibility only for fields in which the graduate is not certified.

For these reasons, the focus of the eligibility questions should be to identify grades and subject fields in which a graduate is eligible to be certified but is not yet certified. The following is a suggested outline for collecting certification and eligibility data that will accomplish this purpose. The different wording needed in part C is easily accomplished with a CATI data collection.

A. ASK WHETHER GRADUATE IS CERTIFIED (Q53)
Do you have any type of certificate or license to teach school at any grade level from prekindergarten through grade 12, in at least one state? That is, are you certified to teach in at least one state?

Yes
No (SKIP TO C)

B. Ask all certification questions: grades, date, kind, state agency, subject fields (Q54-Q57C, Q59-Q60)

C. Ask whether graduate is eligible, using different wording depending on whether graduate is certified (Q50):

IF CERTIFIED: Among the grades and subjects in which you are not certified, are there any in which you are eligible to be certified? By eligible we mean completed all coursework, including student or practice teaching, required for a certificate or license to teach at any or all levels, prekindergarten through grade 12, in at least one state.

IF NOT CERTIFIED: Are you eligible to be certified? By eligible we mean completed all coursework, including student or practice teaching, required for a certificate or license to teach at any or all levels, prekindergarten through grade 12, in at least one state.

Yes
No (SKIP TO NEXT SECTION)

D. Ask all eligibility questions, reworded as necessary to collect only those grades and subject fields in which the graduate is not certified: grades, date, subject fields (Q51, Q52, Q58)

Questions 51 and 54 - Grades Eligible or Certified to Teach

Eliminate the "All Grades" Category. This category was intended to reduce response burden by allowing graduates (or interviewers) to mark 1 category instead of 14. However, the grades that are included in "all grades" is subject to interpretation. Does it include prekindergarten, kindergarten, and ungraded? During data collection, when the "all grades" category was chosen, the CATI system automatically coded "yes" to prekindergarten, kindergarten, grades 1-12, and ungraded. However, the results of the Validity Study indicate that some graduates who were not certified in prekindergarten, kindergarten, or ungraded chose the "all grades" category. It appears that more accurate information can be obtained by asking graduates to indicate exactly which grades they are certified to teach, rather than using the "all grades" category.

Eliminate the "Ungraded" Category. On the survey data file, 471 (unweighted) records have the "ungraded" category in question 54 answered yes. However, 464 of these cases were automatically coded "yes" by the CATI system for graduates who chose the category "all grades." This means that only 7 graduates specifically chose the "ungraded" category. None of the 10 states in the validity study report a certification category of ungraded. It is possible that none of the 51 states actually have an "ungraded" certification category. The few graduates that think they are certified in "ungraded" could be told to choose the grades that correspond with the ages of students they are certified to teach.

Eliminate the "Subject Certified Only" Category. On the survey, the "subject certified" category was meant for those people who were not certified by grade, but only by subject. Interviewers were instructed to probe for grades in which the respondent was certified to teach a specific subject, and only code "subject certified" if a respondent said that he/she was not certified by grade. If "subject certified" was coded, then no other grade could be coded.

During the main survey data collection, only 28 graduates chose this category in question 54. None of the 10 states in the validity study reported certification by subject only and not by grade. It is possible that none of the 51 states actually have a "subject only" certification category as we defined it. The few graduates that think they are certified by "subject only" could be told to choose the grades that their certification allows them to teach, since we are obtaining the subjects in which they are certified in a different series of questions.

Question 56 - Kind of Certificate or License

This question had the highest rate of mismatches between graduate and state reported data on the validity study. The main reason for these mismatches appears to be different interpretations of the reporting categories. None of the 10 states included in the validity study used classifications exactly the same as those used on the survey for this question. In Appendix G, the classification system used by each of the 10 states is examined. This examination reveals that many different terms are used, and the same terms are used in different ways by different states to classify teacher certification. This makes it extremely difficult to develop a standardized system for all states.

For these reasons, NCES should examine the purpose of this question - what information should the question be obtaining and how will this information be used? Data from this question have not been included in previous published reports from the RCG studies. If this question remains in the survey, then certification categories used by each state should be reviewed to develop questionnaire categories that best reflect those used by the state agencies. In particular, NCES should examine whether it is important to make a distinction between initial/provisional and regular/standard certificates, since the difference between these two categories is very ambiguous in some states.

Questions 58 and 59 - Subject Fields Eligible or Certified to Teach

For analysis, the subject eligibility and certification data were used to compare to the subject fields in which the graduate was teaching. The eligibility and certification data were collected first; then employed teachers were asked what subject fields they were teaching in a later section of the questionnaire. The responses were then compared to determine the percentage of teachers who were eligible or certified in their teaching field. Since this is the main purpose of these questions, NCES should consider asking employed teachers directly whether or not they are certified and whether or not they are eligible to be certified for each subject field they are teaching. This would avoid some of the interpretation problems with this question.

Alternately, if the eligibility and certification questions are kept separate from the teaching subject question, some suggested solutions to the interpretation problems are discussed below.

Separate Elementary Fields from Secondary Fields. During the survey data collection, the category "Any elementary fields, general or specialized" was meant to include any respondent certified to teach any subject at the elementary level. Respondents were then expected to answer yes to the specific subject fields only if they had an additional certification in that field. In practice, however, many respondents answered "yes" to each of the subject fields included in their elementary certification, rather than only those in addition to elementary. For example, when graduates with elementary certification were asked whether they were certified in a subject field such as English language arts, they often answered "yes," meaning that they were certified to teach English language arts at the elementary level. Therefore, it is impossible to distinguish between graduates with certification in both English language arts and elementary education, and graduates with certification only in elementary education.

To avoid this problem, elementary certification should be treated separately from other certification. A review of the certification documents for the 10 states included in the validity study shows that the most common certificates issued by these states in the elementary grades are called kindergarten, primary, and elementary education. In addition, some states have certificates in prekindergarten or early childhood education. These prekindergarten certificates should be grouped with the elementary certificates since they sometimes overlap (i.e., prekindergarten/primary). A suggested outline for collecting certification data by subject field follows. Note that by changing the question wording slightly this same outline can be followed for collecting data on subjects in which the graduate is eligible but not certified.

A. Do you have a teaching certificate in any of the following: prekindergarten, kindergarten, primary, or elementary education?

   Yes
   No (SKIP TO C)

B. In addition to your prekindergarten, kindergarten, primary, or elementary certificate, do you have any other teaching certificates or special subject endorsements?

   Yes
   No (SKIP TO next section)

C. I will be reading a list of subject fields. Please tell me in which fields you have a teaching certificate.

IF A=YES, ALSO SAY: Please include only teaching certificates or special subject endorsements that you have in addition to your prekindergarten, kindergarten, primary, or elementary certificate.

LIST OF SUBJECT FIELDS EXCLUDING ELEMENTARY AND PRE-ELEMENTARY

Consider Combining Special Education Categories. For graduates certified in special education, there can be different interpretations of how to fit the certification into the survey categories. For example, a graduate certified to teach "Mildly handicapped K-12," answered "yes" to the specific handicapping conditions that the certificate covers (such as mentally retarded and specific learning disability). However, the state chose the category "General certificate, no specific condition," presumably since no specific condition is named in the certificate. Again, the different interpretations may have been increased since graduates were asked whether they were certified in each of the specific special education fields.

This problem can be dealt with by combining the individual special education categories for analysis. This may also be necessary because the sample sizes for teachers in individual categories may be too small to analyze. If the categories are going to be combined for analysis, then NCES needs to consider whether it is necessary to collect the information by individual category.

Consider Combining Science Categories. There were two problems with the science categories. First, some teachers (especially in junior high or middle school), were certified to teach general science, which was not designated as either biological or physical science. Second, the unweighted sample sizes for graduates employed as teachers in the individual physical science categories (other than general physical science) were very small, ranging from 22 to 49. NCES should consider whether it is necessary to collect this information by individual category. One way to address these problems would be to use the following categories for science:

Any sciences, general or specialized:
- General science (no specialized area)
- Biological or life sciences
- Any physical science (INCLUDE CHEMISTRY, GEOLOGY, EARTH SCIENCE, PHYSICS, AND ANY OTHER PHYSICAL SCIENCE).